Integrated Circuit Design and Technology

TUTORIAL GUIDES IN ELECTRONIC ENGINEERING

Series editors
Professor G.G. Bloodworth, *University of York*
Professor A.P. Dorey, *University of Lancaster*
Professor J.K. Fidler, *University of York*

This series is aimed at first- and second-year undergraduate courses. Each text is complete in itself, although linked with others in the series. Where possible, the trend towards a 'systems' approach is acknowledged, but classical fundamental areas of study have not been excluded. Worked examples feature prominently and indicate, where appropriate, a number of approaches to the same problem.

A format providing marginal notes has been adopted to allow the authors to include ideas and material to support the main text. These notes include references to standard mainstream texts and commentary on the applicability of solution methods, aimed particularly at covering points normally found difficult. Graded problems are provided at the end of each chapter, with answers at the end of the book.

1. Transistor Circuit Techniques: discrete and integrated (2nd edition) — G.J. Ritchie
2. Feedback Circuits and Op. Amps. (2nd edition) — D.H. Horrocks
3. Pascal for Electronic Engineers (2nd edition) — J. Attikiouzel
4. Computers and Microprocessors: components and systems (2nd edition) — A.C. Downton
5. Telecommunication Principles (2nd edition) — J.J. O'Reilly
6. Digital Logic Techniques: principles and practice (2nd edition) — T.J. Stonham
7. Transducers and Interfacing: principles and techniques — B.R. Bannister and D.G. Whitehead
8. Signals and Systems; models and behaviour — M.L. Meade and C.R. Dillon
9. Basic Electromagnetism and its Applications — A.J. Compton
10. Electromagnetism for Electronic Engineers — R.G. Carter
11. Power Electronics — D.A. Bradley
12. Semiconductor Devices: how they work — J.J. Sparkes
13. Electronic Components and Technology: engineering applications — S.J. Sangwine
14. Optoelectronics — J. Watson
15. Control Engineering — C. Bissell
16. Basic Mathematics for Electronic Engineers: models and applications — J.E. Szymanski
17. Software Engineering — D. Ince

Integrated Circuit Design and Technology

M.J. Morant
School of Engineering and Applied Science
University of Durham

CHAPMAN AND HALL

University and Professional Division

LONDON • NEW YORK • TOKYO • MELBOURNE • MADRAS

UK	Chapman and Hall, 11 New Fetter Lane, London EC4P 4EE
USA	Van Nostrand Reinhold, 115 5th Avenue, New York NY10003
JAPAN	Chapman and Hall Japan, Thomson Publishing Japan, Hirakawacho Nemoto Building, 7F, 1-7-11 Hirakawa-cho, Chiyoda-ku, Tokyo 102
AUSTRALIA	Chapman and Hall Australia, Thomas Nelson Australia, 480 La Trobe Street, PO Box 4725, Melbourne 3000
INDIA	Chapman and Hall India, R. Sheshadri, 32 Second Main Road, CIT East, Madras 600 035

First edition 1990

© 1990 Martin J. Morant

Typeset in 10/12pt Times by Best-set Typesetter Ltd, Hong Kong
Printed in Great Britain by Richard Clay Ltd, Bungay, Suffolk

ISBN 0 412 34220 0   0 442 31188 5 (USA)
ISSN 0266–2620

British Library Cataloguing in Publication Data

Morant, M.J.
  Integrated circuit design and technology.
  1. Electronic equipment. Integrated circuits
  I. Title II. Series
  621.381'73
    ISBN 0–412–34220–0

Library of Congress Cataloging-in-Publication Data

available

Contents

Preface

Until a few years ago, all integrated circuits were designed by specialists behind the closed doors of the semiconductor industry, manufactured only in enormous quantities, and sold as standard products. A remarkable change has been brought about by the development of semi-custom design techniques and CAD tools that nowadays enable all electronics engineers to design their own **application-specific integrated circuits** (ASICs) and get them made economically in small quantities. As a result, ASICs have become the key components in electronic products of all types.

This is a book about integrated circuit design and its fundamentals in silicon technology. It is **not** a manual of how to do design using any particular CAD tools, but rather, the background for all of them. IC design started to become an academic subject with the publication in 1980 of the famous book by Mead and Conway which led to the first postgraduate courses in full-custom design in the UK. Since then, ASIC design has rapidly moved into the core of many higher education courses in electronics, aided by the provision of CAD hardware and software that nowadays enable undergraduates to gain practical design experience. With such a rapid development, courses in IC design inevitably contain a rather diverse mixture of computer techniques and parts of more traditional digital and circuit design courses with EA1 and EA2 undertones! It is, of course, all these and more.

ASICs are very often designed using semi-custom methods which require absolutely no knowledge of circuits or silicon. However, the view taken in this book is that the *educated* engineer should understand at least the fundamentals of the circuit and fabrication technologies that have an impact on design and underlie all design decisions. It therefore attempts to fit integrated circuit design into a coherent framework based on the outstanding achievements of silicon technology. IC design provides an excellent opportunity for bringing together interests in semiconductors, digital and analogue circuits, and systems. Although the book tries to be self-contained, it is therefore based on the background of digital and circuit electronics that most students acquire in the first year of a degree or higher certificate course.

The core of the book in Chapters 7–9 describes the essential steps in both semi-custom and full-custom design, and the use and features of the CAD tools required to turn a chip specification into a verified, testable circuit on silicon. The earlier chapters are on IC device structures and how they are made on silicon. Chapter 4 presents the fundamentals of MOS circuits, which can be regarded either as background for semi-custom or the basis of full-custom design. Emphasis is placed on digital CMOS circuits and their design which are most likely to be met in practice, but the book also includes an introduction to analogue CMOS design for mixed analogue–digital circuits. Chapter 5 gives the background of bipolar circuits that will never be completely overshadowed by CMOS.

The book is intended to be an introductory guide to the subject in a rather different sense from other volumes in this Series. It is a guidebook to a country and, for the parts you want to visit, you will need the more detailed information given in CAD manuals, ASIC data books, and more specialized textbooks. In visiting any new country it is useful to know some of the language and I have deliberately introduced some of the established jargon of the semiconductor and CAD industries. I have also had to use centimetre rather than metre units because they are universally used in the industry.

Large parts of this book are descriptive rather than quantitative. It therefore differs from others in the Series in having only a few problems at the ends of chapters. I have omitted problems rather than trying to pretend that short hand calculations can make any real contribution to IC design. Lecturers should have no difficulty in devising CAD projects tailored to the particular hardware and software available and the students' competence in using them. Semiconductor data books can suggest design projects of any difficulty from a few gates to LSI functions, and it is particularly instructive to try to improve on the speed of standard TTL functions by designing them in CMOS. Other design projects will be suggested by courses in communications, computer systems and instrumentation.

My view of IC design, as presented in this book, has developed over the last nine years of teaching the subject in an M.Eng/M.Sc course to which the students and many leading electronics companies have contributed greatly. Too many people have been involved to thank all of them by name, but Simon Johnson has been particularly helpful, not least by providing the well-tried design exercise and some of the figures. I also wish to thank Professor Peter Hicks of the Electronics and Electrical Engineering Department, UMIST, and Professor Tony Dorey, the Consulting Editor, for many constructive comments. Parts of the book were written while I was a Senior Research Fellow at the University of York and I would particularly like to thank the Electronics Department there for their hospitality.

Introduction: Integrated Circuits – Complexity and Design

Reading this chapter should enable you

☐ To get an overview of the main topics that will be developed in the book and an appreciation of the importance of IC design as electronics 'collapses on to silicon'.

☐ To learn some of the terms relating to ICs in general.

☐ To get an appreciation of the immense *complexity* of ICs and the very small dimensions of their components.

☐ To get a feel for the problems that are inherent in IC design, including design cost, accuracy and testability.

☐ To understand the hierarchy of levels in *full-custom* IC design and the need for *semi-custom* methods.

Objectives

In the UK, software for IC design is available to all universities and polytechnics through the national Higher Education Electronics Computer Aided Design (ECAD) Initiative. Many other countries also have facilities for IC design in higher education.

Take the cover off almost any piece of electronic equipment and you will find printed circuit boards almost covered with integrated circuits of many shapes and sizes. What you see are, of course, only the packages protecting the individual silicon chips and connecting them to the board. If you were to open up the packages you could see the chips themselves but you would need a high-power microscope to see the actual circuits on the exposed surface of the silicon.

The first reaction on seeing the very fine detail of an integrated circuit as in Fig. 1.1 is, 'How incredible! How was it made?' or, for those of a scientific frame of mind, 'How does it work?' Engineers should ask, 'How ever was that designed?' This book is intended to provide some answers to that question.

What is an IC?

Before considering design itself we need to be familiar with some of the words used in talking about integrated circuits, or ICs, in general. We should start by explaining just what the term 'integrated circuit' means to systems, design and semiconductor engineers.

We will use 'IC' and the colloquial term 'chip' for 'integrated circuit' throughout this book.

To a systems engineer an integrated circuit is an electronic component used for processing signals in a particular way. Integrated circuits are the building blocks for nearly all electronic systems. Many of them carry out digital functions such as counting or storing digital data, or manipulating it under program control as in a microprocessor. Some, including op amps and many communications circuits, handle analogue rather than digital signals. Others perform both analogue and digital functions, often including A/D and D/A conversion on the chip. To use any type of IC we must be able to understand its electrical specification which will enable us to work out how it will function in a system. The specifi-

An IC on a single chip of silicon is strictly called a **monolithic** IC. **Hybrid** ICs contain several silicon chips and other components interconnected on a ceramic substrate and packaged in an hermetically sealed module for special applications.

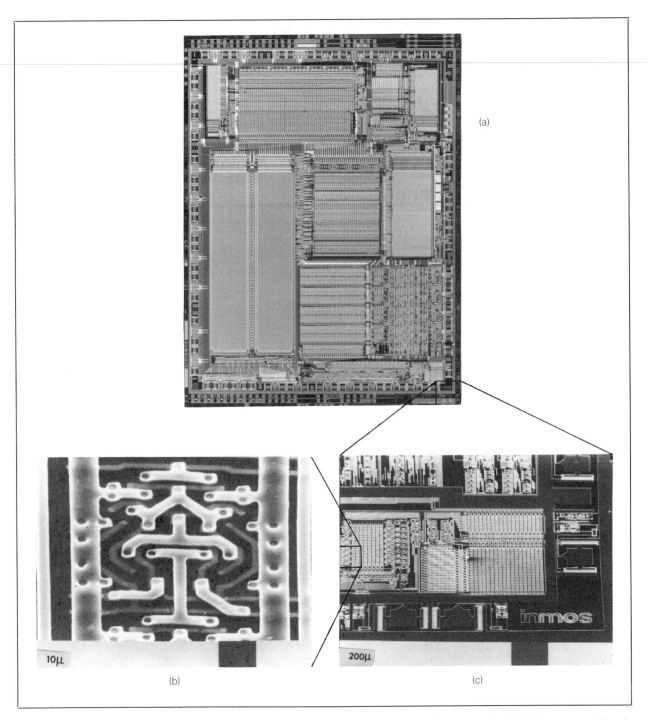

(a)

(b)

(c)

Fig. 1.1 (a) The Inmos T-800 Transputer. This is a truly VLSI integrated circuit made by a 1.2 μm CMOS process on a chip size of 8.5 × 9.5 mm and containing about 400 000 gates. (b) Scanning electron micrograph of the bottom right-hand corner of (a). The marker indicates 200 μm. (c) A further enlargement of part of (b) with the marker now indicating 10 μm. At the scale printed in (c) the entire chip would occupy about 5.3 × 5.9 metres (Courtesy of Inmos Ltd.).

cation, as given in a semiconductor data book for example, is concerned only with the external properties of an IC and this is adequate since its internal operation is largely irrelevant to systems engineers.

To design ICs, however, we need to look inside the package. At this level an IC is an electronic circuit made entirely on the top surface of a thin rectangular piece of single-crystal silicon called a 'die' by the manufacturers or, more colloquially, a 'chip'. The pattern of very small shapes in the various layers on the silicon surface, part of which is shown in Fig. 1.1(b), defines all the transistors and internal connections of a complete circuit. Fairly simple ICs typically contain a few thousand interconnected transistors and the most complex ones a million or more. Each transistor is extremely small and they are packed very closely together on the silicon surface so that a highly complex circuit can be fitted on to a chip only a few millimetres square. Fine wires, welded to contact pads around the edge of the chip, carry signals between the circuit and more substantial connections on the package, and hence to a printed circuit board.

Integrated circuits are manufactured by the semiconductor industry using extremely advanced and exciting 'fabrication technologies' which have their scientific foundations in applied physics and chemistry. A long sequence of fabrication steps is used to build up the layers on the silicon and to etch them into the shapes required to form the transistors and interconnections. Each step has to be precisely controlled to give the best circuit performance and the complete sequence, known as a fabrication **process**, is highly optimized or 'tuned' for the efficient production of a particular class of ICs. The design of an individual chip to be made by this process is contained entirely in the complex pattern of shapes, called the **layout geometry**, that is to be fabricated on the silicon. This is the semiconductor engineer's view of an integrated circuit.

The systems and semiconductor views of an IC are completely different and it is the design engineer who is concerned with the link between them. All design is the process of turning a basic, abstract idea into reality. In this case the idea is the system specification and the product is the piece of silicon and a few other materials which, almost miraculously, implements it as hardware. The design and manufacturing processes bring about this transformation. The designer must therefore know something about how the IC will be used and also about how it will be made, although he is concerned primarily with creating the link between specification and fabrication. There are many levels in this link such as those concerned with algorithms, sub-systems, logic, circuits and layout so that the designer has to take a very broad view of what is meant by an IC. This view is the main concern of this book.

Standard-product ICs and ASICs

Examination of the type numbers printed on the packages of ICs in any piece of modern electronic equipment will show that only some of them are recognizable as devices listed in semiconductor manufacturer's catalogues. These are standard-product or catalogue ICs produced in very large numbers and widely marketed as building blocks for a great range of applications. The others are ICs that have been designed for one particular application and called **application-specific ICs** or ASICs.

Standard ICs are the basic products of the semiconductor industry. They

The term 'chip' is misleading. It always reminds me of a chip of flint! The chips are, in fact, sawn out of the silicon very precisely.

Details of IC packages are given in Sangwine (1987).

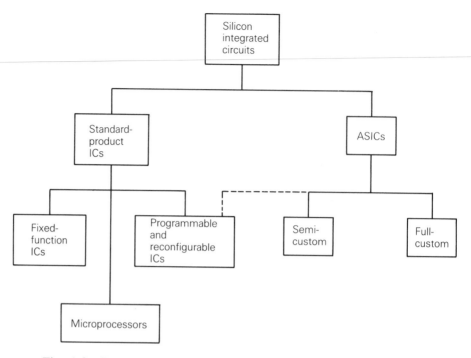

Fig. 1.2   Routes to integrated circuits for building digital systems.

Mass production usually implies quantities of more than a million chips in the production life cycle of a particular design. Memory chips are produced in the largest numbers. One large Japanese manufacturer is reported to be making 10 million, 1 Mbit DRAMs per month at present.

ASICs are usually fixed-function ICs because they are designed for one specific job.

provide the general-purpose digital and analogue functions required in such large numbers by the electronics industry that their cost is kept low by mass production. Many standard ICs are designed to carry out a single fixed function such as digital counting, binary addition, or amplification, although the exact function can often be changed by setting up appropriate control inputs. Standard ICs also include microprocessors which are far more flexible as they can carry out a whole sequence of different functions under the control of a software program, although at a comparatively low speed.

A third group of standard-product ICs are **user-programmable** chips that are electrically programmed after manufacture to carry out a fixed function for a particular application. Many of these can be classed as **programmable logic devices** or PLDs, in spite of the different names used by some manufacturers. Although they are not strictly designed by the user, the ways in which they are programmed have many similarities to ASIC design and we will consider them briefly in Chapter 6.

At first sight it is surprising that there is any need for ASICs in addition to the thousands of types of standard IC described in semiconductor data books. However, one of their main advantages is that a single ASIC may well replace many standard ICs for a particular application. This greatly reduces the cost of manufacturing electronic systems so that ASICs are now commonly used in new products. As a result, most electronic equipment companies now design their own ASICs whereas in the past IC design was entirely the province of the semiconductor industry. This explains why there is now so much interest in IC design in general.

The semiconductor industry is continually bringing out new and improved standard-product ICs made possible by advances in fabrication technology. It takes a great deal of time to design standard-product ICs for maximum performance so that the design cost is extremely high but it is easily recovered when chips are to be produced in enormous numbers. For ASICs however, the design cost is crucially important because of the relatively small number of chips used in the lifespan of a particular piece of equipment to be produced by a single company. The quantity required many only be a few thousand compared with the millions of chips produced for a standard-product design. ASICs therefore only became economical when design costs were reduced by developing new design methods and powerful computer-based design tools. Computer-aided design (CAD) has therefore become as important as fabrication in enabling silicon technology to be widely used. With suitable CAD methods, including **semi-custom** design, ASICs can be designed reliably in far less time than for comparable standard-product ICs, but with some loss of performance. We will have a lot more to say about such methods in Chapter 8.

The cost of design is at least twice as much as the wages of the engineer doing it because of company overheads such as premises and computers. A few weeks' work can therefore add substantially to the total cost of producing perhaps only a few thousand ASICs.

The second problem that had to be overcome before ASICs could become really economical was to reduce the cost of fabricating the relatively small number of chips of the same design that are required for a single application. We will find out more about the cost of producing chips in Chapter 3 where we will see that the low cost of standard-product ICs is partly due to the mass production of each design. For ASICs the cost of fabricating a small number of chips of the same design is reduced by using particular chip architectures, such as **gate arrays**, and other methods that will be described in Chapter 6.

The overall effect of developments in the semiconductor and CAD industries is that the electronics designer now has the wide choice of ways of obtaining ICs for building systems that are summarized in Fig. 1.2. It is important to make the right choice between fixed-function or programmable standard-product ICs, microprocessors, or ASICs for every part of a system. To do this we need a good understanding of integrated circuit technology and of the various ways in which ASICs can be designed.

The Complexity of ICs

Photographs such as those in Fig. 1.1 show that ICs can be extremely complicated. The smallest shapes are associated with transistors and the complexity of an IC can be measured approximately by the number of transistors it contains. Many of the more complex chips are for digital applications and complexity can then be expressed very roughly in terms of an equivalent number of two-input gates, each regarded as four transistors. A **level of integration** can be used to classify ICs by their complexity. Although not precisely defined, the terms in Table 1.1 are commonly used for indicating the complexity of digital ICs. The upper limit of VLSI complexity is still increasing every year but chips containing more than 50 000 gate equivalents are now produced in increasing numbers. They have at least 1000 times as many transistors as in the SSI and MSI chips used in elementary electronics.

A bistable may be taken as equivalent to four or eight gates according to type.

Every transistor in an IC carries out a particular job which contributes in a small way to the overall function. With a large number of transistors acting together, the complete IC can do some very complicated digital processing. The equivalent of a few hundred gates is sufficient for the simpler digital functions

Table 1.1

		Number of gate equivalents/chip
Small-scale integration	SSI	<10
Medium-scale integration	MSI	10–100
Large-scale integration	LSI	100–10 000
Very-large-scale integration	VLSI	>10 000

Unfortunately the term 'VLSI' is commonly used nowadays for far simpler chips.

such as the counting, addition, or multiplication of binary numbers. An eight-bit microprocessor can be made with a few thousand gate equivalents, an advanced 16-bit one requires a few tens of thousands and so on.

Even with only a few thousand gates, the complexity of the chip behaviour becomes nearly as difficult to comprehend as the complexity of the appearance. Complex ICs are best regarded as systems in their own right, and the user may be completely unaware of the individual gates and bistables, let alone the transistors, that contribute to its overall function. The designer, however, has to understand the detail in order to build up all the elements into a complete system. One of the main problems of digital design at this level is in the handling of large amounts of information on both the electrical behaviour and the geometry of the chip. Computer systems are obviously essential for handling this information.

The functions of most analogue integrated circuits such as filters or amplifiers are usually less complex and easier to understand. Analogue ICs tend to use far fewer transistors, rarely more than a few thousand or so, but they swap the complexity of digital circuits with the need for precision in the signal voltages and currents. Analogue design therefore presents rather different problems from digital design and these will be discussed in Chapter 10. Chapters 7–9 will be concerned largely with the design of digital ICs where complexity is nearly always the main problem. These two classes of design are brought together in ICs with both analogue and digital functions on the same chip and we will return briefly to these in Chapter 10.

Figure 1.3 excludes memory chips. The largest memory chips usually contain more transistors than the most complex logic ICs but we exclude them here because their design and production has become a highly specialized branch of the semiconductor industry. Also the regular layout of memory chips enables the transistors to be packed together far more tightly than in general-purpose chips.

The increasing complexity of the most advanced digital ICs available each year has been brought about by continuous refinement of the manufacturing technology. The story of the growth in the number of transistors that can be produced on a single chip of silicon is well known. As shown in Fig. 1.3(a), the earliest ICs in 1961 contained about 10 transistors. By the early 1990s, ICs with more than a million transistors will be commonplace, and a steady increase can be expected well beyond that. Such chips represent the leading edge of microelectronics production technology, but they do not reduce the need for far larger numbers of simpler types, right down to MSI sizes, that are quite adequate for many applications.

The logarithmic annual increase in complexity was first noted in 1964 by G. Moore who later became a founder of Intel Corporation. It has continued ever since then, as predicted, although with a slightly lower slope than in the early years.

In technological terms, the increase in the maximum complexity of ICs has been made possible:

1. By reducing the size of the individual transistors, and
2. By increasing the maximum size of chip that can be manufactured economically.

In design, it has only been possible to cope with this rise in complexity by equally dramatic advances in CAD.

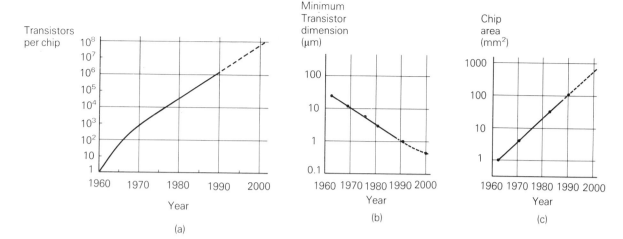

Fig. 1.3 (a) Increase in the maximum complexity of digital ICs from 1961, (b) the minimum feature size of their transistors, and (c) the maximum chip area. Diagrams like these are frequently published but it is not always made clear what type of chips they refer to. Here it is for logic ICs that are in sufficiently large-scale production to be readily available. Memory chips and ICs under development (and sometimes already advertised!) may be considerably more complex.

To appreciate the achievements of fabrication technology we need to have a feel for the dimensions of the transistors and other components in an integrated circuit. The sizes are usually expressed in micrometres or microns. A micron is one thousandth of a millimetre and well below any length we can relate to directly – a hair from your head has a diameter of about 60 μm for example. It is nowadays possible to make parts of transistors with well-controlled widths of one micron or less.

Figure 1.3(b) shows how the dimensions of the smallest transistors in production have fallen as the fabrication technology has been refined. As a result of a tremendous amount of research and development the already-small sizes were reduced by more than 20 times in about 25 years and further reductions are made every year as the technology becomes even more advanced.

Making transistors smaller has two great advantages for ICs. The first is that circuits operate faster as the transistors are scaled down so that some digital ICs for example now work at clock frequencies of at least 100 MHz. The second is that the cost of producing ICs does not rise greatly with the density of transistors, that is the number per square millimetre, that they contain. By shrinking the dimensions, the cost per transistor therefore falls rapidly. With smaller dimensions, closer packing, and other technological improvements the cost of transistors on large ICs did, in fact, fall by about 10^4 times in the first 25 years of silicon chip production. The continuing increase in packing density also explains how it is possible to produce more and more complex chips each year with only small increases in cost. Reducing the transistor size has been the largest single factor in bringing about the explosive growth in the use of electronics which has been called the Information Revolution. The trend will certainly continue for many more years before fundamental limits to the minimum size are eventually reached.

The micrometre (10^{-6} m = 10^{-4} cm) is called a **micron** in the semiconductor business. The smallest particle visible to the naked eye with good illumination has a diameter of about 40 μm.

The electrical characteristics change as transistors are made smaller. A point will be reached where they become either unusable or subject to unacceptably large statistical variations, but the limits have not yet been identified with certainty.

The second development leading to even more complex digital ICs has been the steady increase in the maximum size of chips in large-scale production which is shown in Fig. 1.3(c). How this has come about will be explained in Chapter 3 when we consider some of the economic aspects of chip fabrication that are important for design. We will find that, in general, the cost of chip fabrication in large-scale production rises only slowly with complexity so that the cost per gate is reduced dramatically by increasing the level of integration. This is the stimulus driving the semiconductor industry to greater and greater complexity. For the systems designer it means that it is often cheaper to use a single complex IC rather than several smaller ones.

One of the problems that had to be faced by the semiconductor industry some years ago was that, while it was possible to mass produce more and more complex chips, it became increasingly difficult to find general-purpose functions that were likely to be required in sufficiently large numbers to bring the prices down. As complexity rises the functions produced become more specialized and demand falls. The first large uses for LSI were therefore in new applications such as calculators and watches. The demand for even more complexity, and eventually VLSI, was later created by the development of standard-product microprocessors that could be mass-produced and sold in very large numbers because of the wide range of applications made possible by software programming. More recently, large-scale specialized applications have been found for standard-product VLSI in communications, computers and signal processing, although the need for general-purpose building blocks is falling with the increasing use of ASICs.

These have been called **application-specific standard-product** (ASSP) ICs.

The overall effect of developments in silicon fabrication technology is that the ever more complex ICs produced each year lead to even more powerful information systems, opening up entirely new application areas for electronics. However, although electronics itself can be said to be collapsing on to silicon, the increasing complexity does raise some really challenging technical problems, particularly, as we shall see, in the area of IC design.

The Problems of Design

In order to get the most benefit from silicon fabrication technology we must be able to design ICs efficiently and quickly. Nearly all the problems in designing digital ICs are ultimately due to complexity. The first, and probably most important problem, is how to reduce the time taken in designing complex circuits. It is obvious that we cannot design a circuit containing say 20 000 transistors one transistor at a time! Even if it was technically feasible, it would be prohibitively slow and expensive so that we have to find better ways of dealing with the complexity.

We have already seen that the time spent on design is crucial because its cost can add appreciably to the final cost of the chips even when moderately large numbers are to be produced. For example, it might cost £50 000 to design a standard IC containing 5000 gates, and this would add 20p to the cost of each chip for a production run of 250 000. If the design cost was proportional to complexity it could rise to £1M for a similar IC containing 100 000 gates and this would have to be covered by a far larger production run. A low and predictable design time is particularly important for ASICs where it may account for a large proportion of the total cost.

Design cost depends very much on the complexity of the chip function. It is certainly possible to design a 5000-gate ASIC for far less than £10 000 if its function is well understood at the start.

The other problems that have to be overcome in IC design are concerned with accuracy, testability, and optimization of the performance. 'Good' design should lead to the production of a chip to satisfy four requirements.

1. It should function correctly and meet the specification when it is fabricated for the first time.
2. It should have the expected electrical performance in speed and power.
3. It should be capable of being thoroughly and rapidly tested in production.
4. It should use the silicon technology efficiently to reduce the chip size and hence the production cost as far as possible.

Accuracy is especially important in IC design because even a single error in the final layout can make a chip useless. It can be very difficult to locate errors and it may take a great deal of extra time and cost to correct them. Errors can creep in at any stage in the design process and it is not always possible to check or verify that a design is completely correct before fabrication. Even though the final design may be thoroughly checked, there are so many possible sources of error that it would be rash for the designer to claim that it is 100% accurate before fabrication. However, good design methods, care, and attention to detail can certainly give a high probability that the final IC will function correctly and at the expected speed.

An additional problem for the design process is that ICs are totally integrated, and not assembled from separate parts. In designing other engineering systems it is usually possible to check that some if not all of the parts work correctly by building and testing them before assembly. For example, a bread-board prototype of a circuit is often built to check that the design is correct before committing it to a printed circuit board, and all its individual ICs will certainly be tested before connecting them together. However in designing the ICs themselves, it is usually not possible to make practical measurements on individual parts because of the prohibitive cost and delay that would be involved in fabricating a special test chip just to reassure the designer! The entire design of an IC therefore has to be completed as a single intellectual task in the mind, on paper, and in the computer, before any of it is made.

Because an IC is so fully integrated as a single complex product the amount of information needed to describe its operation and structure expands enormously during the design process. The specification of a simple IC might be as short as half a page of text but its description will eventually expand to the final layout data which is many orders of magnitude larger. The layout data for a digital IC with the equivalent of 5000 gates, for example, will contain the dimensions and position coordinates for about 100 000 separate shapes. Luckily, this amount of data can easily be handled by quite low-cost computer systems and design would be impossible without them.

All the problems of IC design increase with complexity, particularly the need to reduce the time spent on design without sacrificing quality. The problems are not in the amount of data as such but in how to make the best use of computers and the computer aided design (CAD) tools that might be available. To overcome the problems careful consideration must therefore be given to the overall design strategy or **methodology** even for the design of ICs containing only a few thousand gates which is the maximum that many readers are likely to encounter in practice. We will return to the far greater problems of large VLSI design in Chapter 11.

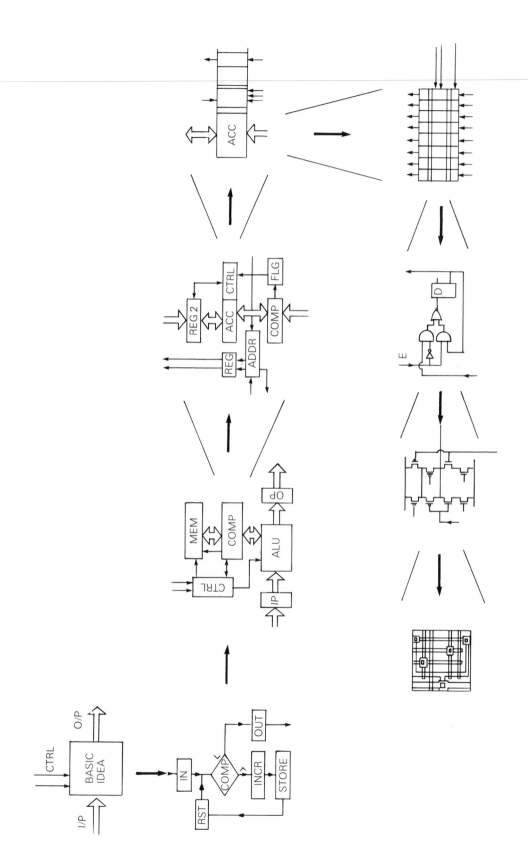

Fig. 1.4 The hierarchy of levels in IC design illustrated by the stages in designing a special-purpose comparator chip.

Hierarchical Design and CAD

The problems inherent in the design of digital ICs can be greatly reduced by adopting a methodical, hierarchical approach with extensive use of the appropriate CAD tools. **Hierarchy** means splitting a problem down into a series of levels each of which is completed before proceeding to the next. For the design of a complex digital IC we can identify levels concerned with the following.

1. Specification: The overall chip function and its precise specification as in a data sheet.
2. Functional operation: Methods of implementing the function expressed initially as an algorithm and finally as a hardware block diagram.
3. Partitioning to register level: The splitting-up of the blocks by stages until recognizable digital functions such as counters, registers and combinational logic are reached.
4. Gate-level logic: The logic design of the digital functions in terms of cells containing gates and bistables.
5. Transistor circuits: The design of circuits for each of the cells.
6. Chip layout: The geometry of the shapes to be etched in the layers on the silicon to form the transistors and circuits.

These are the levels for design as shown diagramatically in Fig. 1.4. However, an integrated circuit is not an end in itself as it is just a component for use at higher levels. It is usually mounted with other chips on a printed circuit board to form a self-contained sub-system. The board will then be connected to others through the backplane of an enclosure making a single item of electronic equipment which may still be only a small part of the final system. An IC designer must understand the immediate use for the chip and also know something about its eventual applications.

The lower end of the hierarchy also extends below the design of chips to the design of transistors, and to fabrication, both of which are closely related to the basic physics of semiconductors. One of the attractions of microelectronics as a subject is this complete and coherent framework extending all the way from the fundamental science and technology of semiconductors to the operation of complete electronic systems. However, the range is so wide that nobody can understand all of it in depth, and it is necessary to specialize on a few of the levels, which here will be those at which the IC designer actually works.

'Top-down' hierarchical design starts from a basic idea, 'We want a chip to do....' The requirements are put together into a precise specification which fully describes the system function and the algorithms and procedures by which the output will be generated. This is an abstract, text-based, 'behavioural' description of the chip function and in the design process it is translated by stages into 'structural' descriptions of interconnected hardware blocks. This translation starts at the top level by constructing the highest level 'block diagram' for the entire chip, showing interconnected blocks such as input/output units, logic processors and memory. The design then proceeds by considering the behaviour of each block and how it can be made from interconnected sub-blocks which themselves are made up of smaller blocks, and so on.

Top-down design continues through as many stages as may be needed down to the gate, transistor, and layout levels. At each stage the design is expanded and it is essential to make sure that its overall operation is the same as in the previous level. This is done wherever possible by using computer simulation to verify that every level of design is correct before proceeding to the next. This is the only way of producing a reliable design in a reasonably short period of time, although, as we will find when considering the details in Chapters 8–9, top-down design is not necessarily as straightforward in practice as it appears to be.

We will find that one of the main skills in hierarchical design is in partitioning the functions so that the final blocks will contain only a small number of different basic cells combined in regular repeating arrays. Structures of this type can greatly reduce the design time and improve the chances of the chip working correctly the first time.

Every level of IC design places great reliance on the CAD tools that are used. Accurate simulation of the behaviour of interconnected blocks, logic and transistor circuits is essential because of the impossibility of practical verification at the end of each stage. Modern ECAD software is extremely reliable but it is still only a tool and it must be used in the right way to aid and not control the designer's creativity.

ECAD = Electronics CAD, to distinguish it from other engineering CAD software.

Semi-custom and Full-custom Design

Application-specific ICs only became economic with the development of **semi-custom** design methods which reduce the design time by shielding the designer from all the details of circuit and fabrication technologies. They do this by supplying complete designs for basic **cells** which are commonly used circuit elements such as various types of gates, bistables and larger blocks from which all digital functions (and in some systems analogue ones also) can be built. The data on these pre-designed cells is contained in a **cell library** in the CAD system. Semi-custom design is therefore analogous to the design of systems using standard ICs except that it is done entirely within the CAD system.

With semi-custom methods, the designer does not need to know anything about the internal details of the cells and, as he is not allowed to modify them either, he is completely isolated from the circuit and layout levels of design. The final design of a semi-custom IC contains only the positions of the cells and their interconnections and it is expanded to the actual layout by the semiconductor manufacturer. This greatly reduces the design time and eliminates errors where they are most likely to occur. It also reduces the amount of knowledge needed by the designer and enables digital and systems engineers to design ICs correctly without necessarily understanding circuits, semiconductors or fabrication.

Design styles that do go down to the circuit and layout levels are known as **full-custom** design. Specially designed circuits can be optimized to operate faster and to be fitted into a more compact layout than the general-purpose cells used in semi-custom methods, but at the expense of a greatly increased design time and a greater likelihood of errors. For full-custom design the semiconductor manufacturer provides the electrical parameters and layout design rules for a particular fabrication process, but not the details of the production methods, so that design is still well separated from fabrication.

It is only in research and development departments within the semiconductor industry itself that a designer might be able to interact with the highly tuned fabrication processes to improve the performance of transistors or circuits. Even here though, the designer's freedom of action will be limited by the company's

The **electrical parameters** for a fabrication process include such things as the thicknesses and electrical properties of the layers forming the transistors and other components. They are used for calculating the properties of transistors and circuits in computer simulations.

The **layout design rules** are the geometrical constraints on the shapes of the individual layers. They include, for example, the minimum widths of conducting tracks, and the overlaps that must be allowed between edges on different layers to ensure reliable fabrication.

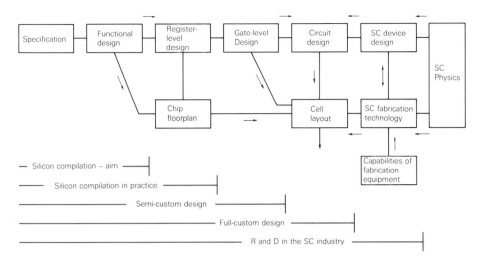

Fig. 1.5 The range of levels available in various IC design styles.

technology and ultimately by the basic physics of fabrication processes and of semiconductors.

We will have a great deal to say about semi-custom and full-custom styles of design and their comparison in Chapters 8–9. Figure 1.5 shows the levels at which a designer may actually work. They can be divided into:

1. Functional and logic design, and
2. Circuit and cell design.

A 'top-down' style of design proceeds from left to right. The design levels indicated by the five blocks on the left are common to semi- and full-custom design. Full-custom continues to cell and circuit design. The four blocks on the right-hand side of Fig. 1.5 are never available to the designer.

A third style of design known as **silicon compilation** is also shown in Fig. 1.5. This aims at removing the designer from all the detailed levels of the chip operation by automatically converting a description of the chip function, written in a high-level language, into layout. Most of the present silicon compilers convert structural descriptions of chip function into layout but the long-term objective of much current research is to produce VLSI layout for complete chips directly from behavioural descriptions. We will briefly return to this in Chapter 11.

Some existing silicon compilers already work from behavioural descriptions for small parts of digital ICs such as PLAs, but to do this for a complete complex chip function is far more difficult.

If any design problem is to be split up between designer and manufacturer, there must be a well-defined interface for exchanging information between them. For semi-custom design the manufacturer must provide a complete cell library and at least advice on CAD and, for full-custom design, the design rules and the electrical properties of the layers and transistors. The designer must also work within specified guidelines to produce a design submitted in a specified data format for fabrication.

The CAD tools used for design are often aimed at either semi- or full-custom design styles. Many semiconductor companies provide semi-custom CAD systems for their own products, but it is also possible to use general-purpose tools with the addition of a cell library provided by a manufacturer for semi-custom design.

The choice between full- and semi-custom design styles is determined in industry by company priorities. For companies producing electronic equipment containing ASICs, the priorities are usually to get working chips as quickly as possible in order to bring out a new product. The success of a company can easily depend on the introduction of a product ahead of the competition and delay in designing a key element such as an ASIC can be disastrous. Semi-custom design is comparatively quick and reliable and the reduced chip performance may not matter. If performance is important a full-custom ASIC might be preferred but it would take a far greater and less predictable amount of time to design.

Priorities are different in the semiconductor industry where design is largely concerned with standard-product ICs for applications that need high performance and minimum costs for large quantities. This usually necessitates full-custom design, the cost of which is far less important for a mass-produced chip.

By limiting the number of levels for design, engineers can now be trained or retrained specifically in IC design skills, usually for semi-custom digital design using a particular CAD system. It is perfectly possible to design very good ICs without knowing anything about fabrication, devices or even circuits. However, 'the educated designer' should have a wider perspective including some knowledge of the IC technologies that underpin all design styles. We attempt to give that perspective in Chapters 2–5.

Summary

This chapter has introduced many of the main aspects of IC design that will be developed further in this book. It has attempted to give an appreciation of the magnitude of the problems inherent in IC design caused by the immense complexity of function and circuitry that can be achieved in a single chip of silicon. The problems of design require a hierarchical top-down approach, starting from a precise specification and working through functional and logic design to circuits and layout. Advanced CAD tools are essential for every level of design to handle the complexity and to verify that the eventual chips will function correctly.

IC design has to be targeted at a particular circuit technology and fabrication process provided by a semiconductor company and the designer cannot deviate from the rules associated with the process. The design overheads can add appreciably to the cost of producing chips. For the relatively small number of chips required for an ASIC design, the cost is greatly reduced by using semi-custom methods based on libraries of predefined cells which avoid the need for design at the circuit and cell layout levels.

Problem

If a gate circuit forming part of an IC occupies an area of $20L \times 20L$, where L is the minimum dimension of a transistor, calculate the number of gate equivalents the might be fitted into (a) a 5×5 mm chip, and (b) a 15×15 mm chip if $L = 2\,\mu m$ and if $L = 0.7\,\mu m$.

Integrated Circuit Components and Structures

2

Studying this chapter should enable you

□ To describe the physical structures of the components commonly used in ICs from the point of view of design.
□ To learn some of the terminology on wafer fabrication that is needed for design.
□ To get an appreciation of why a designer has to accept the electrical properties provided by a fabrication process.
□ To understand the results of doping semiconductors and the formation of p–n junctions.
□ To be able to draw the shapes of MOS and bipolar transistor structures modified to isolate them electrically in ICs.
□ To understand why other components used in ICs are often based on transistors or parts of transistors.

Objectives

Technology and Design

In this and the following three chapters we will consider those aspects of silicon integrated circuit technology that are important for design. In technology we will include fabrication, device design, circuit topology and the electrical level of circuit operation. Design will include the digital and systems levels that will be described in Chapters 7–9.

We should first get a clear idea of what the digital IC designer really needs to know about the extensive subjects of devices, circuits, and fabrication. The answer depends very much on the type of design being undertaken.

As we saw in Chapter 1, semi-custom methods for designing ASICs can be highly efficient because they impose a barrier between digital and circuit design which saves the great deal of time needed for working at the lower levels. A semi-custom designer can therefore get by with very little detailed knowledge of chip technology. What is needed is an appreciation of the advantages and disadvantages of various types of ASICs for building digital hardware, and sufficient experience to make the correct fundamental design decisions, for example:

1. The choice of semi- or full-custom ICs or of PLDs for a particular design,
2. The choice of circuit technology, manufacturing process and supplier,
3. The estimation of the likely chip size and cost.

It requires at least a basic understanding of circuits and of the economics of chip fabrication to make these important decisions and to evaluate manufacturers' claims for performance and cost. However, after choosing a suitable process and CAD system (a 'route to silicon'), the technology becomes hidden within the cell library if a semi-custom method is adopted. This has parallels in many other branches of engineering where the designer has to accept the properties of materials and components without question in order to concentrate on the more creative levels of design.

Full-custom design, on the other hand, requires a much deeper understanding of circuits to achieve the highest possible speed and perhaps to consider using novel circuit forms. This is even more true of the design of analogue ICs and of cell libraries. To use the flexibility of full-custom design, the electrical and device parameters for the chosen fabrication process have to be known, but these are supplied by the semiconductor manufacturer. Even the full-custom designer cannot work below the circuit level in designing ASICs so that strictly he needs to know only a little about the production details which actually determine the electrical properties of the layers.

Established fabrication processes are standardized within every semiconductor company because of the very large investment in 'tuning' them for maximum efficiency. Designers are not, therefore, allowed to use circuits that would require changes to be made in fabrication processes. If the characteristics of the transistors and other components that can be designed using the available layers are not good enough, the designer has no alternative but to seek a more advanced process elsewhere. The only exception to this is in the semiconductor industry itself where designers might be able to influence the development of new device and fabrication processes for improved circuit performance and this needs a deep understanding of chip electronics at all levels.

Most of this chapter is concerned with the shapes and sizes of the components that we need to design, and their fabrication will be outlined in Chapter 3. Chapters 4 and 5 describe the electrical properties of the devices and circuits used in semi-custom ASIC design. More advanced circuits that are used in full-custom design will be described in Chapter 10.

Silicon Technology

Silicon is a semiconductor, one of many materials with rather peculiar electrical properties that are used in ICs primarily to make transistors. Because of its importance in electronics it has been studied scientifically in far greater detail than any other material. It has a unique and almost ideal combination of physical and chemical properties that enable us to fabricate high-performance transistors and ICs at very low cost. Only one other material, the compound semiconductor gallium arsenide (GaAs), has any advantages over silicon for making transistors and then only for devices for use in very high-speed digital and microwave circuits. However, it is far more difficult to fabricate monolithic ICs in GaAs than in silicon so that it is only used for very specialized applications where the greater cost can be justified. This book is concerned specifically with silicon ICs although the principles of design are the same for any semiconductor.

The manufacture of all silicon ICs uses developments of the **planar** process introduced about 1960 for the production of silicon transistors. The word 'planar' refers to the production of device layers on the flat top surface of the chip even though modern ICs, with many more layers, are far from flat microscopically, as shown in Fig. 1.1(c).

A more important feature of the planar process is that a large number of chips are produced simultaneously on a circular **wafer** of silicon. This is a slice less than a millimetre thick cut from a cylindrical single crystal of silicon which can

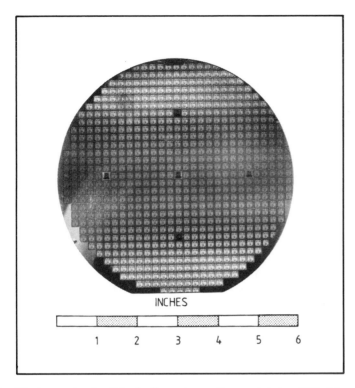

INCHES

| | 1 | 2 | 3 | 4 | 5 | 6 |

Fig. 2.1 Photograph of a 6 inch diameter wafer containing 672 chips. The five distinctive chips ('drop-ins') contain structures for making electrical measurements to check that the processing has been done correctly. The flat edge at the bottom of the wafer is accurately aligned along one of the crystal axes for use as a reference. (By courtesy of Plessey Semiconductors Ltd.)

nowadays be grown to a diameter of six or even eight inches. Many hundreds of chips can be fabricated on a single wafer which, when complete, is cut up to separate the individual chips for packaging. Figure 2.1 shows a 6 inch diameter wafer before cutting. The production of batches of 20–30 wafers simultaneously, each of them containing hundreds of chips, is the key to the low cost of ICs.

Integrated circuits contain 10 or more electrically distinct layers on the top surface of the chip. Some of the layers are formed inside the silicon by **doping** it in selected areas with certain elements, and others are thin films of either conducting or insulating materials added to the top surface. The conducting layers are **patterned** into complex shapes in plan view, that is looking on to the silicon surface. Some of the smaller shapes may be parts of transistors, while others form conducting tracks for interconnections. The insulating films separate the conducting layers except where contact holes are etched through them to allow connections to be made between the different layers for 'wiring-up' the circuit.

The edges of some of the shapes and the contact holes are essentially what we see by using an electron microscope as in Fig. 1.1(b). A three-dimensional view, Fig. 1.1(c), is obtained by increasing the magnification and this shows some of the overlapping shapes of the different layers much more clearly. It can be seen that the layers are even thinner than their plan-view dimensions and they are usually between 40 and 1000 nanometres in thickness.

The combination of methods used for the efficient production of ICs by the

Yes, inch units are commonly used for wafer diameters and die sizes in the semiconductor industry.

It is a pity that this book is not printed in colour. Under a microscope, chips often appear to have beautiful pastel colours caused by interference in the thin transparent films on the surface.

1 nanometre = 10^{-9} m = 10^{-3} μm. Very roughly atoms are about 0.1 nm diameter so that a 40 nm thick film has about 400 atomic layers.

semiconductor industry makes it one of the greatest technological achievements of all time. It takes many hundreds of highly controlled wafer processing steps to form the layers and to etch them to the required shapes. Extreme accuracy, control and cleanliness must be maintained at each step. In particular, all the layers must be very pure and of the correct thicknesses. The dimensions of the shapes have to be precisely controlled and their relative positions on the different layers have to be accurate to within a fraction of a micrometre right across the wafer in order to control the overlap of the layers which creates the thousands of transistors and connections on each chip. As the technology has firm foundations in physics and chemistry, every aspect of fabrication has been subjected to close scientific study and continuous technological refinement which is still going on. Its future depends on the further development of automated production equipment, very special clean rooms, exceptionally pure chemicals and gases, and many other specialized items developed by supporting industries.

Clean rooms are essential to reduce contamination by airborne particles. These are not the particles you see in a beam of sunlight but the far larger number of microscopic particles present in air. Particles down to 0.2 μm diameter are filtered out of clean-room air. Gases and chemicals have to be similarly filtered.

The sequence of steps, every detail of how they are carried out, and the production equipment used defines a wafer fabrication process. This is a 'recipe' for making a particular class of ICs and it is often specific to a single semiconductor company or even to a particular fabrication plant. The names used to describe processes often give the minimum width of a shape and the type of circuit, e.g. 'a 1.2 μm double-metal n-well CMOS process'. The significance of these terms will become apparent in Chapter 4.

Every fabrication process is highly developed by a semiconductor company to give the largest possible output of working ICs with the best performance. It requires a very considerable investment in equipment and time to set up a new process and to determine the device characteristics needed for design. The process will be used for the production of many different designs and the manufacturer therefore tries to keep it stable for several years. This explains why designers have to work within the constraints of a particular fabrication process although it may be possible for a design to be transferred from one process to a similar one with only small changes.

The final design of any IC is completely defined by:

1. The layout, giving the geometry of the shapes on each of the layers, and
2. The full details of the fabrication process which determine the electrical properties of the layers, the characteristics of the transistors and hence the electrical performance.

The ultimate output of the design process is therefore the layout data for the chip which is necessarily contained in a computer file. This is the starting point for its fabrication on silicon.

Integrated Circuit Components

Before considering fabrication processes in more detail we need to know something about the components of ICs and the physical structures that we have to design and make. Transistors are by far the most important components in ICs because they are the active, control elements essential for all electronic circuits. They are also the most critical components to produce, requiring close dimensional tolerances and silicon of extremely high purity and crystalline perfection in order to obtain useful and reproducible electrical characteristics. In comparison, the resistors required for some circuit forms have to take second place as we will see later.

Transistors can be made with extremely small dimensions without affecting their electrical characteristics and some properties actually improve as they are scaled down. Reducing the dimensions allows a larger number of transistors to be packed on to a given area of silicon so that more complex circuits can be produced without the price rising. The minimum size of transistors is determined by the smallest dimensions that can be made reliably by the fabrication process used and every new process introduced by the industry includes a further reduction in device dimensions.

Transistors are devices that enable the current in one circuit to be controlled by the voltage in another. As the control action is very sensitive they can produce amplification or high-speed digital switching depending on the type of circuit they are used in. Transistors make use of the special properties of electrons in semiconductors, in this case silicon, to bring about this control action. How they work is quite an involved story that is not particularly important for design. IC designers have to accept the properties of transistors offered by a fabrication process, although in full-custom design some adjustment can be made by changing the dimensions of the transistors. There are, however, a few basic facts about the properties of semiconductors and the physical structures of the components that must be known by all IC designers.

The operation of p–n junctions and transistors is fully described by Sparkes (1987).

n-type and p-type Silicon

The electrical conduction of crystalline semiconductors such as silicon depends critically on chemical purity. An absolutely pure or intrinsic crystal of silicon, if it could be made, would be a fairly good insulator at room temperature. For making ICs the starting material is very highly purified silicon but it is always doped with small quantities of another element when a crystal is grown.

The elements used for doping silicon are from Groups III and V of the periodic table, on either side of Group IV which contains silicon itself. Group V elements such as phosphorus, antimony or arsenic are known as **donors** when they are used to dope silicon because each donor atom provides the silicon with an extra electron which is free to move through the crystal and carry current when a voltage is applied. Silicon doped with donors is said to be **n-type** because of the negative charge of the added electrons which increase the conductivity.

The alternative is to dope the silicon with a Group III element, typically boron, which is known as an **acceptor** in silicon because each atom extracts one of the valence electrons linking the silicon atoms in the crystal. The vacancy left behind can move from atom to atom through the crystal, behaving very much like a free electron contributed by a donor but with a positive charge. The moving vacancy is known as a **positive hole** and it can be thought of as having nearly all the properties of a free electron except for the opposite charge. Silicon doped with acceptor atoms is therefore said to be **p-type** (p for positive).

If a voltage is applied to a p-type sample the holes move through the crystal as positive charges, drifting towards the most negative potential. The p-type doping therefore increases the conduction in much the same way as n-type doping but with the important difference that the sign of the charge carriers is reversed.

Extremely small concentrations of donors or acceptors are used in doping silicon. The basic crystals are lightly doped, perhaps with as little as one part in a

The resistivity of intrinsic silicon is about 250 000 Ω cm. This is 2500 Ω m in SI units but cm units are always used in the semiconductor industry instead of metres.

A 'free' electron is one that is not attached parmanently to an atom but it is free to move around through the crystal. Without an applied field the electrons move around at random due to their thermal energies. With a field, the drift velocity is superimposed on the random motion.

The movement of electrons and holes is in opposite directions when a voltage is applied to pieces of n- and p-type silicon. The conventional current direction is the same in both cases because of the opposite signs of the charge carriers.

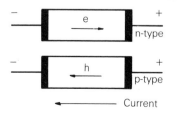

19

hundred million of either type of element. During the fabrication process, larger concentrations of donors or acceptors are added to small regions on the top surface of the silicon to form the transistors. When the silicon is doped with more than about one part in 10^5 of a donor or acceptor element it is said to be heavily doped n^+ or p^+.

The p−n Junction

The difference between p- and n-type semiconductors is not particularly significant in itself for ICs. However, when a p−n junction is made we have the most important device on which the whole of modern electronics depends. A junction can be made by adding acceptors, for example, to a small region of a piece of n-type silicon as shown in Fig. 2.2(a). If we also make metal contacts to the p- and n-type regions so that a voltage can be applied, the p−n junction is found to be an extremely efficient electrical rectifier with a d.c. characteristic as in Fig. 2.2(b). With the p-type positive and the n-type negative, defining the forward direction of the diode, the junction can pass a large current with only a small voltage drop, while with the opposite polarity, the reverse direction, the current is extremely small up to tens or hundreds of volts.

Integrated circuits contain many thousands, if not millions of p−n junctions and we will find out later more about how they are made. A typical junction might have an area of $10 \times 10\,\mu m$. In the forward direction a current of $100\,\mu A$ will give a voltage drop of about $0.65\,V$, but in the reverse direction the current is often no more than $1\,pA$ up to 10 or $20\,V$.

The rectification property of a p−n junction is due to the formation of a potential barrier between the p- and n-type regions. The n-type silicon contains conduction electrons and the p-type contains holes. This difference sets up a contact potential, typically about $0.6\,V$, between them, which is present even when there is no externally applied voltage. The potential difference occurs across a very thin layer extending for a small distance on both sides of the actual junction. Nearly all the electrons and holes are pushed out of this layer by the electric field so that it is depleted of charge carriers and is hence called a **depletion layer**. The width of the layer increases with the reverse voltage but it is typically about $1\,\mu m$ or less, and it determines many of the dimensions of transistors.

Depletion layers are extremely important in semiconductor devices and ICs of

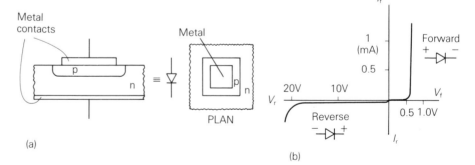

Fig. 2.2 (a) Cross section of a p−n junction made in silicon by adding p-type dopant over a small area of the top surface. Plan view is to reduced scale. (b) Direct-current current−voltage characteristic of a p−n junction.

all types. Many of the p–n junctions in ICs are reverse biased. This increases the potential between the p- and n-type regions pushing out the electrons and holes even more so that very few of them pass through the depletion layer and the current is negligibly small. The depletion layer therefore behaves like an insulating layer embedded in the junction and separating the p- and n-type regions electrically. This property is used extensively in ICs to electrically isolate one piece of silicon from another.

Although the d.c. current of a reverse-biased p–n junction is negligible in an IC, the depletion layer has capacitance like any other insulating layer and this becomes extremely important at high frequencies. Appendix 1 describes some of the important features of depletion layer capacitance.

When a p–n junction is forward biased the potential across the depletion layer is reduced so that the electrons and holes are less constrained to their respective n- and p-type regions. As a result carriers of both types can move through the depletion layer giving the relatively large forward current.

Bipolar and Unipolar Transistors

Transistors are three-terminal current-control devices. When suitable bias voltages are applied the current that flows between two of the terminals is controlled by the voltage applied to the third connection. There are two quite different ways in which the control action can be achieved in transistor structures and they give rise to the two main classes of devices which are:

1. Unipolar transistors in which the current is carried by one type of charge carrier only, and
2. Bipolar transistors in which the current is carried by both electrons and holes.

It is very remarkable that both classes of transistor are widely used in ICs of dif-

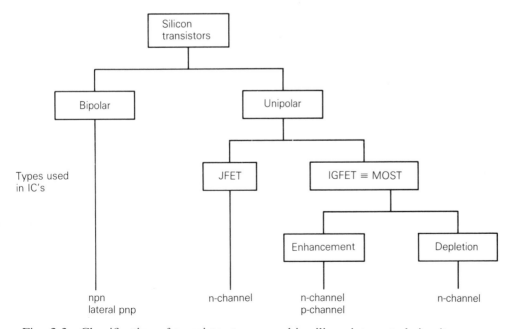

Fig. 2.3   Classification of transistor types used in silicon integrated circuits.

ferent types and that their properties have continued to be comparable throughout the recent development of semiconductor technology. Whether this is an indication that they are not as distinct as they appear to be is a philosophical question that we will not pursue here.

A further subdivision of silicon transistor types is shown in Fig. 2.3. In unipolar devices the control action uses the electric field due to the input voltage but in somewhat different ways in junction field-effect transistors (JFETs) and in insulated-gate field-effect transistors (IGFETs). Silicon JFETs are not used in digital ICs, and although they have a specialized role as input devices in analogue amplifier circuits, we will not need to consider them here. IGFETs, on the other hand, in the form of metal-oxide-semiconductor transistors (MOSTs or MOSFETs), are the main devices in more than three quarters of the ICs in use today.

The two types of transistor give rise to the two main classes of ICs – MOS and bipolar. Both have advantages and disadvantages which the designer must understand in order to decide which type should be used for a particular application. We are concerned here with their physical shapes and sizes and will consider their electrical properties in Chapters 4 and 5.

The Basic MOS Transistor Structure

The word 'structure' is used for the physical layers and shapes that make up a device.

The gate oxide thickness has to be less than 100 nm to get useful current control in an MOS transistor. It is typically less than 40 nm thick in modern devices and its thickness is sometimes expressed in Angstrom units (Å). 1 Å = 0.1 nm.

The basic n-channel MOS transistor has the structure shown in Fig. 2.4(a). It uses a p-type substrate in which two small rectangular areas are converted to n-type. Metallic contacts to these areas form the **source** and the **drain** connections, S and D. The control electrode is the **gate**, G, which is a conducting layer, often made by depositing a thin film of silicon on top of an extremely thin insulating layer of silicon oxide, the **gate oxide**. The device is classed as an IGFET because there is no direct connection between the gate and the silicon. This is emphasized by the circuit symbol for the MOST.

One would not normally expect any current to flow when a voltage of either polarity is applied between source and drain because the structure contains two p–n junctions back-to-back and one of them must always be reverse biased. However, current does flow when a sufficiently large positive voltage is applied to the gate, because a conducting channel is formed by electrons induced into the surface of the silicon. The current carried by the flow of these electrons between the source and the drain is controlled by the voltage applied to the gate. The term 'n-channel' refers to the negative sign of the charge carriers.

The critical dimension in the fabrication of MOS transistors is the channel length L between the source and drain junctions (Fig. 2.4(a)). It is a plan-view dimension, as seen looking down on to the silicon, so that it is determined by the accuracy with which shapes can be formed on the surface. At the present time, channel lengths of between 1 and 3 μm are normal on ICs.

Another type of MOS transistor is the p-channel enhancement MOST shown in Fig. 2.4(b). This is geometrically the same as the n-channel device but the type of doping in all three regions is reversed. The main current is now carried by the flow of holes in a channel between source and drain as controlled by a negative gate voltage, the term 'p-channel' referring to the positive sign of the charge carriers. This type of MOST operates with circuit voltages of the opposite polarity from the n-channel device which is why they are said to be complementary to each other.

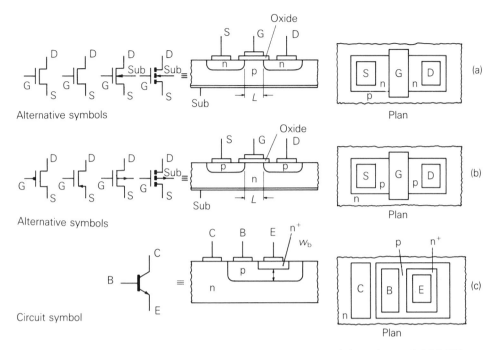

Fig. 2.4 Schematic diagrams of transistor structures (a) n-channel MOST, (b) p-channel MOST, (c) bipolar, n–p–n. (Plan views to reduced scale.)

The Basic Bipolar Transistor Structure

The most common type of bipolar transistor has the n–p–n structure shown in Fig. 2.4(c). The two p–n junctions are known as the **emitter** and **collector** junctions, with connections E and C, which correspond to the source and drain of the MOST. The control electrode is the common p-type region, called the **base**, with its external connection B. The main current in the bipolar transistor is carried by the flow of electrons vertically through the extremely thin base region between the emitter and collector junctions. It is controlled very sensitively by the voltage applied between the base and emitter, B and E, by the process of **bipolar transistor action** which is quite different from the control action in the MOST. The importance of the p–n junctions in the operation of this device is emphasized by calling it a **bipolar junction transistor** or BJT.

We will have more to say about transistor action in Chapter 5.

    There are two aspects of the design of bipolar transistors that are particularly important for fabrication. The first is that the dopant concentration has to be very accurately controlled. In particular, the n-type silicon forming the emitter junction has to be heavily doped which is why it is shown as n^+ in Fig. 2.4(c), while the collector needs a relatively low doping concentration. The second is that the distance through the base between the emitter and collector junctions has to be extremely small for the best high-frequency performance. This distance, known as the base width w_b, is typically less than 1 μm. Both the thickness and doping of the base layer have to be exceptionally well controlled in the fabrication of bipolar transistors.

    Figure 2.4(c) is a diagram of an idealized bipolar transistor in order to show the concept and, as we shall shortly see, it has to be changed quite a lot to make a real transistor for an IC.

Device Structures for ICs

The diagrams in Fig. 2.4 show only the main features of device structures for ICs. They have to be modified substantially in designing transistors to have the best possible electrical performance, to include electrical isolation, and to allow for the practicalities of fabrication. We will describe the more complicated physical structures used in typical modern ICs, omitting the many small variations made by individual semiconductor companies.

A very large number of transistors and other components are made in the single piece of silicon forming an integrated circuit, and it is important to ensure that they do not interact electrically except through the circuit connections. The entire circuit is embedded in a piece of silicon called the **substrate** and it is not immediately apparent how to prevent it from being shorted out by conduction through the semiconductor.

Electrical isolation between components is not only possible but it is achieved in two quite different ways in modern ICs:

1. By using reverse-biased p–n junctions, and
2. By forming oxide barriers on the top surface of the chip.

Although these methods provide d.c. isolation, they both introduce capacitance between the components that must be allowed for in the design of high-frequency circuits.

Isolated MOS Transistors

One of the main advantages of the MOS transistor is that it is inherently isolated from the silicon substrate by p–n junctions if it is correctly biased in a circuit. Isolation is achieved if the source–substrate junction of every MOSFET is either shorted or reverse biased so that its depletion layer is effectively insulating, preventing current flow to the substrate. The drain region of every MOSFET is similarly isolated because the polarity of the circuit voltage always makes its junction reverse biased. The conducting channel, being in between the source and drain, also forms a reverse-biased junction to the substrate with negligible d.c. leakage.

The sources and drains are not always labelled on the cross sections of MOS transistors because they have to be defined by reference to the circuit as the devices are symmetrical.

Many MOS transistors can therefore be made on a single piece of silicon without leakage currents between them, provided that the potential of the substrate keeps the source regions at zero or reverse bias. However, if a metal connection passes over the silicon in between two of the transistors, as in Fig. 2.5(a), an unwanted **parasitic** MOS transistor is formed between them. If this became switched on inadvertently the leakage would disrupt the operation of the entire circuit and this is prevented nowadays by growing a comparatively thick oxide layer, known as the **field oxide**, in between the MOSFETs to isolate them laterally (Fig. 2.5(b)). This also has the added advantage of reducing the capacitive coupling between the transistors.

The thick oxide increases the voltage required to turn on the parasitic MOST to a value above the normal operating voltage of the circuit.

The field oxide is grown by heating the silicon in an oxidizing atmosphere. As the silicon oxide (SiO_2) is formed, it consumes silicon from below so that it grows downwards as well as upwards relative to the original surface. The growth of the field oxide has to be localized to selected regions around each transistor, to form the isolating barriers shown in Fig. 2.5(b). The transistors themselves

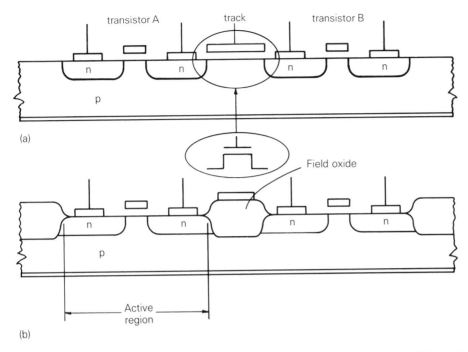

(a)

(b)

Field oxide

Active region

Fig. 2.5   The formation of a parasitic MOST between transistors A and B and its suppression by the addition of an oxide isolation barrier (b).

are made in the relatively small un-oxidized, or active, regions on the surface. All modern MOS production processes use oxide isolation with trade names containing words such as Iso- (isolated with oxide).

The section and plan view of a real MOS transistor with metallic tracks leading to the source, drain, and gate, are shown in Fig. 2.6. The connections from the metal to the device are made through contact windows that are etched in the thin layer of oxide that covers the entire circuit. Figure 4.4 (page 52) attempts to show the MOST in three dimensions.

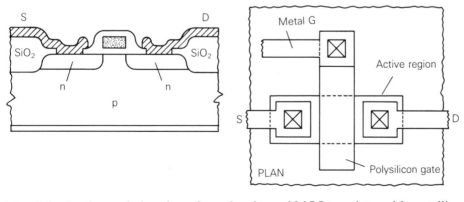

This is the first figure that attempts to show the cross section of a real device. The vertical dimensions in such diagrams have to be exaggerated to show the very thin layers clearly. For example the source and drain junctions in Fig. 2.6 are at a depth of less than a micron compared with a sideways dimension of several microns. The back of the wafer say 270 μm below the top surface would be one metre below the bottom of the page if it was drawn to the same scale as the layer thicknesses!

Fig. 2.6   Section and plan view of a real n-channel MOS transistor with metallic connections to source, drain and gate. The plan view dimensions could be as small as about 5 μm square.

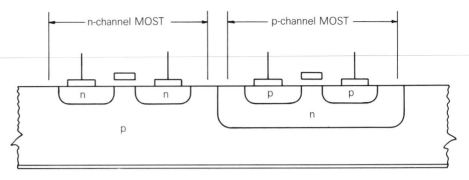

Fig. 2.7 The principle of n-well CMOS.

CMOS Structures

There are many advantages in using MOS circuits containing both n- and p-channel transistors. These are called **complementary MOS**, or **CMOS**, circuits. As shown in Fig. 2.4, n- and p-channel MOS transistors are made in silicon of the opposite polarities, p- and n-type, respectively. To make both types of transistor in a single piece of silicon we therefore need areas of each polarity on the top surface of the chip. Starting with, say, a p-type substrate, selected areas, called **n-wells**, are therefore doped n-type. Groups of n-channel MOSTs are made in the p-type areas, and p-channel MOSTs are made in the n-wells (Fig. 2.7). The wells form p−n junctions to the substrate which, providing that they do not become forward biased, isolate the regions in which the two types of transistor are made.

This type of CMOS fabrication is called an n-well process. An alternative is to use a p-well in an n-type substrate. Both n-well and p-well processes are widely used and the choice between them depends on the type of circuit to be employed, as we shall see in Chapter 4.

The cross-section of a pair of transistors formed by a typical n-well CMOS process is shown in Fig. 2.8. There are many variations on this type of structure produced by different manufacturers. One alternative that can improve the electrical performance is to use both p- and n-type wells in a near-intrinsic epitaxial layer although, with more fabrication stages, this is a more expensive technology.

Another variant is to use **silicon-on-insulator**, or SOI CMOS technology (Fig. 2.9) in which the silicon substrate is replaced by a wafer of a crystalline insulator such as synthetic sapphire. The transistors are fabricated in a film of silicon

A simplified p-well structure looks like this.

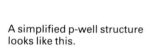

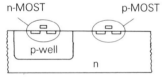

The wells of this 'twin-well' process would look like this. n-channel transistors are made in the p-well and vice versa.

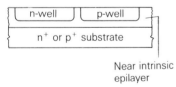

Near intrinsic epilayer

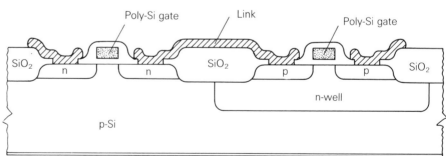

Fig. 2.8 Section of a complementary pair of MOSTs forming part of a CMOS IC. They are shown connected together with a link as in an inverter circuit. The horizontal distance could be as small as about 15 μm.

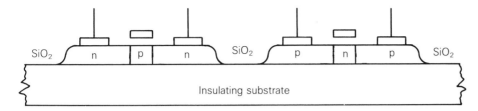

Fig. 2.9 The principle of silicon on insulator CMOS.

grown epitaxially on the insulator. This is a special-purpose technology developed largely to make ICs that are less affected by nuclear particle radiation and it is still extremely expensive to produce.

In yet another technology, BiCMOS, that we will consider further in Chapter 5, bipolar transistors are added to a basic CMOS process to improve the electrical performance.

Isolated Bipolar Transistors

The simple bipolar transistor structure in Fig. 2.4(c) cannot be used in an integrated circuit because the collector is connected to the substrate. Two transistors made on the same piece of silicon in this way would therefore have their collectors connected together which would not be very useful for making circuits! The electrical isolation of BJTs therefore requires some rather substantial changes to the structure.

The first step in modifying the BJT structure is to add a p–n junction below the transistor to isolate its collector from the substrate (Fig. 2.10(a)). For an n–p–n transistor a p-type substrate is now used and the transistor itself is formed in an n-type layer on the surface. The chip design has to ensure that the p–n junction between the substrate and the n-type layer is always reverse biased to isolate the transistor vertically. As the active part of the transistor has to be made in crystalline silicon, the n-type layer must maintain the crystal structure of the substrate when it is grown. The growth process for forming a layer in this way is known as **epitaxy** and the n-type material is said to be an **epitaxial layer**, which is shortened colloquially to **epilayer**.

The second step in modifying BJTs for integrated circuits is to isolate them in the sideways or lateral direction. This is done in modern ICs by surrounding each transistor with an oxide barrier that cuts right through the n-type epilayer (Fig. 2.10(b)). As in MOS, the field oxide covers most of the surface and the transistors are made in the unoxidized active regions. Transistors made by these methods go under trade names such as LOCOS (locally oxidized silicon) or Isoplanar.

Unfortunately, the transistors of Fig. 2.10(b) would have poor electrical characteristics due to the high resistance in series with the collector. The collector current is carried by electrons flowing across the junction below the emitter and then through the high resistance of the n-type epilayer to reach the external collector contact. To reduce this resistance two highly conducting n^+ layers have to be added (Fig. 2.11). The first is a buried n^+ layer added below the transistor to reduce the lateral resistance. It is formed in the p-type substrate before the epitaxial layer is grown and it has no direct electrical connection to the external circuit. The second extra layer needed to reduce the collector series resistance even more is an added n^+ region below the contact itself. This is often made at

Bipolar transistors have to be made in single crystal silicon to get a high and reproducible gain.

Note, however, that the function of the field oxide barrier is to provide d.c. isolation for a bipolar transistor in addition to eliminating parasitic transistor effects that are its main reason in MOS.

27

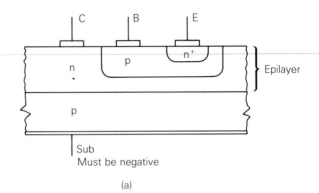

(a)

The large zig-zags at the sides of the p-type substrate indicate that it is far thicker than the epilayer. The ratio of the thicknesses is about 100 : 1 in practice.

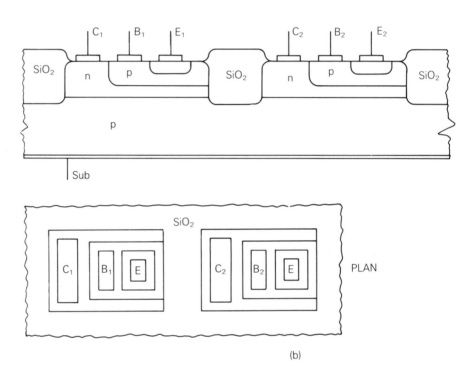

(b)

Fig. 2.10 (a) The principle of junction isolation used to make an n–p–n bipolar transistor electrically isolated from the substrate, and (b) two completely isolated n–p–n transistors. (Plan view to reduced scale.)

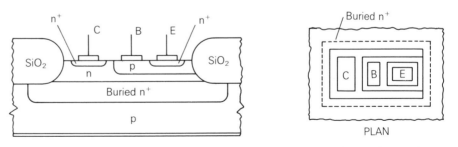

Fig. 2.11 Full structure of an isolated n–p–n bipolar transistor. (Plan view to reduced scale.)

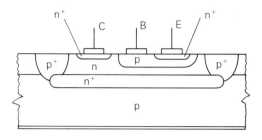

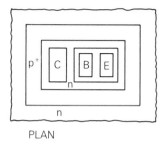

PLAN

Fig. 2.12   An n–p–n transistor structure with junction isolation in the lateral as well as the vertical direction. (Plan view to reduced scale).

the same time as the n^+ emitter, although in some processes it is an additional layer that is made to go right through the n-type epilayer to contact the buried layer below and reduce the series resistance even further.

An alternative to oxide isolation is to surround each transistor with a ring of p^+ doping extending right down to the substrate and providing junction isolation laterally as well as vertically (Fig. 2.12). This was, and still is used in older processes where the base layer is too deep for oxide isolation. Because extra allowances have to be made for manufacturing tolerances, junction isolated transistors are far larger than the more modern types.

The structure of the bipolar transistors used in many modern ICs contains some further modifications in order to reduce the size (Fig. 2.13). The first is that the field oxide is sometimes arranged to separate the base and collector regions of the transistor to prevent them interacting when they are very close together. The second is in the plan view where, with oxide isolation, the base and emitter regions can extend right across the active part of the surface, i.e. the part that is not covered by field oxide. Another advantage of the oxide barriers is that the metal contacts to the base and collector regions do not have to be so accurately defined where they pass over the oxide. Transistors made by these methods can be smaller than $10 \times 20\,\mu m$ including half the width of the oxide surround.

Another method of lateral isolation, developed more recently, is 'trench' isolation. In this, a trench is cut in the silicon surrounding each transistor by a selective etching technique rather than oxidation. A thin insulating layer of oxide is then grown on the inside walls of the trench and it is refilled with

Oxide barriers can be made narrower than p–n junctions for lateral isolation. They also have a lower capacitance which is necessary for high-frequency operation.

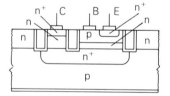

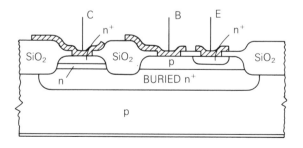

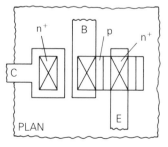

PLAN

Fig. 2.13   Structure of a real oxide-isolated n–p–n bipolar transistor. (Plan view to reduced scale). Possible dimensions go down to about $10\,\mu m$ square.

polysilicon to make the surface flat enough for tracks to pass over the top. The advantage of trench isolation is that it occupies less area, but it is more expensive because of the more complex fabrication technology. Trench isolation is also used in some advanced MOS processes.

Interconnect, Contacts and Vias

The connections between the components of an IC can be made at various levels as in a multi-layer printed circuit board. The main connections are made in conducting tracks formed by the patterning of metal films deposited on the surface. The tracks must be insulated from the silicon except at contacts such as the three terminal connections on transistors. The insulation is usually provided by another layer of silicon oxide which is an extremely good electrical insulator. The contacts are made by etching holes in the silicon oxide before the metal is deposited. These holes are called **contact windows**.

In MOS circuits, the gate layer, often formed by a thin deposited **polysilicon** film, is also used for making interconnection tracks in a similar way to metal but for short distances only because of the higher track resistance in this layer. The diffused layers in the silicon are also used for interconnections between components over shorter distances still.

Circuits can be made smaller if the metal connections can be made on two or more levels. In a **double-metal** fabrication process two layers of metallization are used with a polyimide insulating layer in between (Fig. 2.14). Contacts are made between the layers by etching holes, called **vias**, in the polyimide before depositing and patterning the second layer of metal. Some manufacturers also have processes with two layers of polysilicon to make the interconnect more compact.

Resistors for ICs

Silicon technology concentrates on the fabrication of small high-performance transistors because they are the most important elements in any circuit. In the case of MOS its success has led to circuits that use transistors entirely, although some of them may be connected to carry out the functions of resistors. However bipolar circuits generally require some resistors that need to be approximately linear so that they cannot be made from transistor structures as such.

Resistors for bipolar integrated circuits generally use the same layers as the bipolar transistors on the same chip because extra layers are very expensive to

'Polysilicon' is short for 'polycrystalline silicon'. It is the form taken up by silicon when it is deposited on to SiO_2. It is made up of extremely small randomly oriented crystals, or **crystallites**, of silicon.

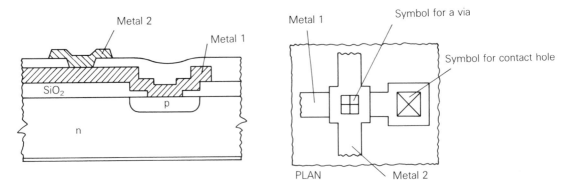

Fig. 2.14   Use of two layers of interconnect on ICs. (Plan view not to scale.)

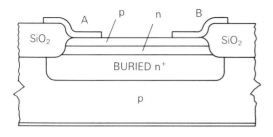
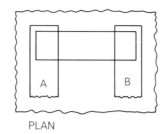

PLAN

Fig. 2.15 A resistor for an oxide isolated bipolar IC. The metal connections are *A* and *B*. The buried n^+-layer is included to suppress parasitic transistor action. (Plan view to reduced scale.)

fabricate. A resistor can, for example, be made from a rectangular p-type strip formed in an n-type layer with contacts at each end (Fig. 2.15). The p-type layer forming the base of n–p–n transistors has a suitable resistivity and it is often used for making such resistors. It is isolated vertically by the p–n junction which must be reverse biased in the circuit, and laterally by an oxide barrier as used around the transistors.

Resistors of this type generally have values limited to about 10 kΩ if they are not to take up an excessive area. They have poor electrical characteristics and tolerances that have to be allowed for in circuit design. If it is essential to use more precise resistors they can be made in polysilicon films deposited on top of silicon oxide but this requires more elaborate processing and it is only used for special purposes.

Other Components

A few additional components are needed for certain circuit applications and again they are usually made using the same layers as the transistors. Only three need to be mentioned here.

1. Diodes are usually made from transistor structures of either type by connecting the base or gate to one of the other terminals.
2. Capacitors may be made from reverse-biased p–n junctions in which case the capacitance varies with voltage (Appendix 1). Better, parallel plate capacitors for analogue circuits are made using polysilicon-diffusion, polysilicon-metal or two layers of either polysilicon or metal.
3. Bipolar p–n–p transistors are usually made as **lateral** devices with the current flow parallel to the silicon surface between emitter and collector regions that are separated sideways. This structure can be fabricated at the same time as the normal n–p–n transistors.

The p–n–p transistor is a rather specialized component used primarily in bipolar analogue circuits and it is unlikely to be required for ASICs unless one of the circuit technologies based on **integrated injection logic** (I^2L) is used.

Summary

Silicon integrated circuits are made from 10 or more electrically distinct layers, some of which are n- or p-type doping layers in the silicon while others are insulating and conducting films formed on the top surface of the chip. The conducting layers are **patterned** into complex shapes in plan view and holes are etched at precise positions in the insulating films to make contacts between electrically separate layers. The design of an IC is completely defined by the layout geometry of the shapes and the details of the fabrication process for producing a particular type of circuit.

The p–n junction is the basic device in all modern electronics and the thickness of the depletion layer at the junction ultimately determines the minimum plan view dimensions of transistors which are of the order of 1 μm. This chapter has described the structures containing p–n junctions and other layers that have to be designed and fabricated to form MOS and bipolar transistors and complete ICs. The problem of electrically isolating transistors made in a common semi-conducting substrate is overcome by using reverse-biased junctions and oxide barriers grown into the silicon surface. Complete circuits are formed by adding connections at various levels.

Problems

2.1 Draw to scale the cross section of the gate oxide of an n-channel MOST with a channel length of 1 μm and an oxide thickness of 35 nm.

2.2 The drain region of an MOST has plan-view dimensions of 20×20 μm and a thickness of 0.5 μm. How many atoms does it contain if silicon has 5×10^{22} atoms per cm^3.

2.3 If the drain region is doped with donors at a concentration of 1 part in 10^7, how many donor atoms does it contain?

2.4 Draw the plan view of two MOSTs connected in series without an external circuit connection between them.

2.5 Draw the plan view of an npn transistor with its collector connected to a resistor made of a p-doped layer and with a contact to a metal connection between them.

2.6 Calculate the capacitance of the p–n junction between the drain region of the MOST in Problem 2.2 and the substrate if there is 5 V between them. Use the information given in Appendix 1 with $C_{j0} = 0.4$ fF μm^{-2}, $\psi = 0.63$ V, and n = 0.5.

2.7 Draw Fig. 2.8 to scale given the following dimensions:

Channel lengths	2.0 μm
Source and drain lengths	8.0 μm
Contact windows	2.5 μm square
Junction depth	0.5 μm
Gate oxide thickness	35 nm
Field oxide thickness	1.1 μm
Polysilicon thickness	600 nm
Metal thickness	800 nm
Well depth	3.0 μm
Substrate thickness	0.3 mm

Integrated Circuit Fabrication Processes 3

Studying this chapter should enable you
- ☐ To understand which aspects of integrated circuit fabrication have an impact on design.
- ☐ To understand the basic processes of doping, oxide masking, photolithography, and metallization used in all silicon device fabrication, and the additions required for many IC fabrication processes.
- ☐ To learn how the individual fabrication stages are used to build up a complete CMOS circuit.
- ☐ To appreciate the requirements for photolithographic masks for ICs.
- ☐ To learn about yield, the cost of producing chips, and the effects of chip area on the cost.

Basic Fabrication Processes

To illustrate the methods used in fabricating ICs we will start by considering how the p–n junction diode, Fig. 2.2 (reproduced a bit more realistically in Fig. 3.1), can be made. We will assume that a diode of this size and type is to be made on every chip that will be fabricated simultaneously on a single wafer of silicon. The diode uses an n-type substrate and we need to make a square p-type region in the same position on every chip. We therefore have to consider how a dopant such as boron can be added to silicon, how to control the area over which it is added, and how to form the metal contacts.

Diffusion and Ion Implantation

Dopant atoms can be added to silicon in either of two ways:

1. By diffusion, or
2. By ion implantation.

Diffusion involves heating the silicon to a high temperature in an atmosphere containing a compound of the element to be added, in this case boron (Fig. 3.2(a)). At high temperatures the compound decomposes and boron atoms are deposited on the silicon. Some of these atoms diffuse into the surface by substituting for silicon atoms in the crystal lattice and, with continued heating, they move a short distance into the silicon. When the wafer is cooled this thin surface layer remains doped with boron, making it p-type. The distribution of boron atoms with depth from the surface can be accurately controlled by varying the time, temperature, and atmosphere for diffusion. Typically, temperatures of about 1000 °C and times of an hour or so are required for boron to diffuse to a depth of about one micron.

Ion implantation uses a beam of high-energy ions, in this case B^+, directed on

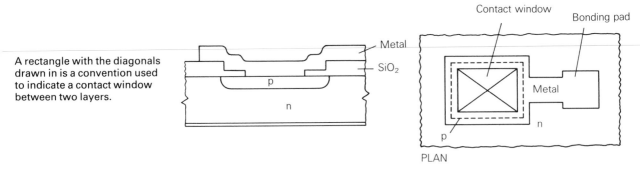

A rectangle with the diagonals drawn in is a convention used to indicate a contact window between two layers.

Fig. 3.1 A practical p–n junction diode. The bonding pad is for the attachment of an external wire connection. (Plan view to reduced scale.)

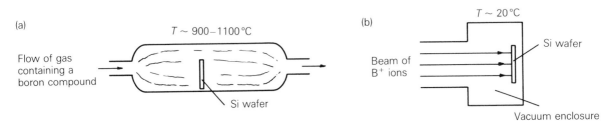

Fig. 3.2 Methods of doping silicon with boron: (a) diffusion and (b) ion implantation.

An ion implanter is a very large piece of equipment costing in the order of £1M.

to the silicon in a vacuum chamber (Fig. 3.2(b)). The ions come to rest a short, well-controlled distance below the surface but they only become electrically active when the silicon is subsequently annealed at about 600 °C. Energies of between about 30 and 100 keV are used to implant the silicon to depths of up to about a micron from the surface. Ion implantation gives better control of the dopant distribution than diffusion and it is used to produce all the shallow junctions in modern ICs.

Whichever method is used, the concentration of the added dopant is greatest at, or just below, the surface of the silicon as shown in Fig. 3.3(a). The distribution of the added dopant with depth is called the **doping profile**. In the present case the acceptor concentration N_a cm^{-3}, added to the top surface is greater than the donor concentration N_d cm^{-3} of the original silicon so that the surface layer is p-type overall. The junction depth x_j is the depth at which N_a and N_d are equal and below this the silicon remains n-type.

The effective doping can be taken as ($N_d - N_a$) or the other way round depending on which is the larger. This is because donors can 'compensate' for acceptors in silicon and vice versa.

Oxide Masking

As described above, the boron would be added to the silicon across the entire area of the wafer. To make a small diode on each chip we therefore have to prevent the boron atoms or ions reaching the silicon except over the square areas of the p–n junctions that we want to make. This is done by using a layer of silicon oxide as a mask (Fig. 3.3(b)). If the doping is by ion implantation, an oxide film of suitable thickness can be used to stop all the ions before they reach the silicon except where square holes or windows are etched out to form the diodes. The oxide mask is just as effective if the doping is by diffusion because

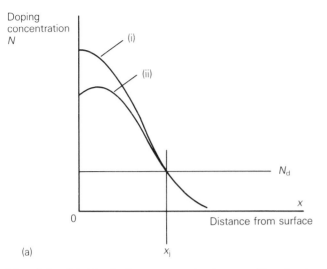

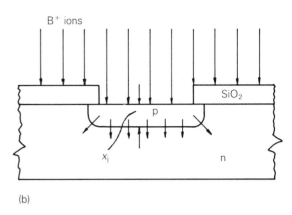

Fig. 3.3 (a) Typical acceptor doping profile obtained (i) by diffusion, and (ii) by ion implantation. (b) Use of an oxide mask for selective doping.

the elements used for doping silicon hardly diffuse at all through silicon oxide at the temperatures used.

Silicon oxide is readily grown on silicon by heating in either pure oxygen or in an atmosphere containing water vapour. The maximum thickness of oxide required for masking is about 1.5 μm and this is usually grown in a few hours in a **wet-oxidation** furnace at about 1000 °C. After growing a layer of masking oxide over the entire wafer, the windows are etched in the diode areas in each chip position. This is done by a process known as **photolithography** which is the basis of the methods for patterning of all the layers of an IC.

The atmosphere in a wet-oxidation furnace contains water vapour in addition to oxygen. This oxidizes the silicon more quickly than if pure oxygen is used.

Photolithography

To make the array of diode chips covering the silicon wafer the pattern of black shapes shown in Fig. 3.4 must be etched into the masking oxide. The pattern is transferred to the wafer using a process which is like an extremely refined form of photographic printing. The pattern of Fig. 3.4 is first etched into a thin film of chromium on a glass plate which is used as a mask for 'printing' on to the oxidized silicon. In the present case the pattern is very simple and the chromium will be removed from the plate everywhere except in the array of small squares.

The sequence of operations in transferring the pattern to the silicon is shown in outline in Fig. 3.5. The oxidized wafer is first coated with a thin film of an organic photosensitive polymer known as **photoresist**. The mask is brought close to the wafer and the pattern is projected on to the photoresist by exposure to ultraviolet light. The exposed areas are hardened by the radiation, but the dark squares remain unaffected. In the process of 'developing' the image the unhardened parts of the photoresist are removed, uncovering the surface of the oxide in the windows. The wafer is then chemically etched in an acid which removes the oxide from the areas that are not protected by the hardened photoresist, to form the windows in the oxide for the p-type doping. The hardened resist is finally removed by decomposing it in a plasma and the wafer is ready for ion implantation or diffusion.

As described, this is a negative photoresist process in which the exposed part of the resist layer is hardened by the illumination so that it is not removed. An alternative would be to use a positive resist which works the other way around, being softened by illumination. To make the same diode using positive resist the mask would be **dark field**, which is a photographic reversal of the mask for negative resist.

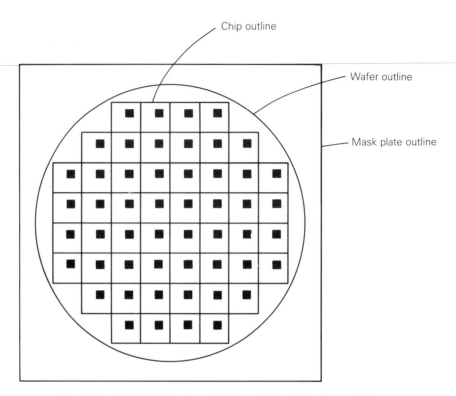

Chip outline

Wafer outline

Mask plate outline

Fig. 3.4 Mask plate for defining diode diffusion windows.

The chemistry involved in photolithography and etching is extremely complex. Wet etching uses very pure acid solutions but dry plasma-etching (in a low-pressure discharge in suitable vapours) is more commonly used where very finely resolved shapes are required. In either case the etching has to be highly selective, removing one material, in this case oxide, but not attacking others, here silicon and photoresist, so as to leave sharp and well defined edges.

Metallization

To complete the diode after forming the p–n junction, metal contacts are made to the small p-type regions and to the back of the wafer. A thin oxide layer will probably have been grown after the p-type layer has been formed and a contact window has to be cut in it to make the connection to the top surface of the diode. This requires a second mask with an array of rather smaller squares than for the doping windows (Fig. 3.6(b)) but used in exactly the same way as in the first stage of lithography. A new problem is that the second mask has to be aligned precisely relative to the first pattern so that the contact holes are accurately placed in the centre of the p-type regions. To get accurate registration, special alignment marks are therefore included in the chip patterns and very sophisticated mask alignment and exposure machines are used for transferring the patterns to the wafers.

The contact metal is usually aluminium containing a small percentage of silicon and other elements. A film of metal about one micron thick is deposited uniformly across the wafer, Fig. 3.6(c), filling in the contact holes and covering the edges of

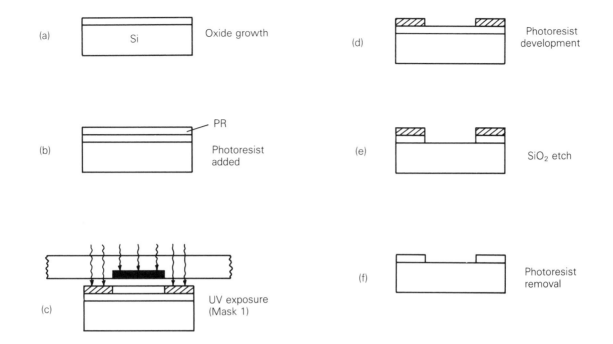

Fig. 3.5 Steps in defining a window in an oxide film on silicon by photolithography.

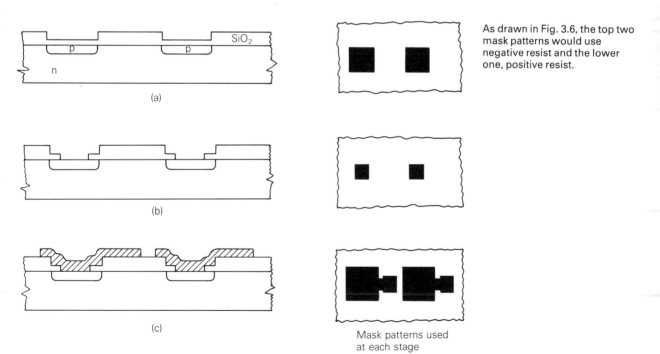

Fig. 3.6 Steps in making the metallic connections to two of the diodes on a wafer, (a) section after doping and re-oxidizing, (b) after cutting the contact windows, and (c) after metallization.

the oxide steps elsewhere. Deposition is by vacuum evaporation or sputtering in a low pressure argon discharge. The metal is then etched into the shape required for the connections which here would be squares somewhat bigger than the contact holes with tracks climbing away over the oxide to bonding pads for wire connections to the package, as in Fig. 3.1. The patterning of the metal uses a third photolithographic mask, which must be accurately aligned to the previous patterns, and suitable chemicals for removing the aluminium cleanly without affecting the oxide.

The back of the wafer is usually protected by oxide during processing. To make the back contact the oxide is removed and a suitable metal alloy film is deposited over the entire wafer. For a low-resistance ohmic contact a metal such as gold containing a small amount of antimony might be used.

Integrated Circuit Fabrication Processes

The steps described above for making a diode illustrate many of the principles used in the fabrication of ICs. They are selective doping by oxide masking and diffusion or ion implantation, etching of contact holes, and the patterning of thin deposited films. A long and complicated sequence of steps is needed even to make a diode and many practical details have been left out of the description to keep it as simple as possible. To make ICs, the sequence is far longer because many more layers have to be made. Our consideration of the structures of IC

Table 3.1   The layers used in bipolar and MOS integrated circuits

	Bipolar	MOS
Layers formed by doping the silicon	Buried n^+ layer	n- or p-well for CMOS
	Epitaxial layer, generally n-type	Epitaxial layer (sometimes)
	Transistor bases, p-type	Sources and drains, n- and p-type
	Transistor emitters, n^+ p-type isolation (sometimes)	
Insulating layers	Field oxide for isolation	Field oxide for isolation
	Oxides between interconnection layers	Gate oxide
	Polyimide between interconnection layers	Oxides between interconnection layers
	Overglaze for protection	Polyimide between interconnection layers
		Overglaze for protection
Conducting layers	1, 2 or more layers of metallization	Polysilicon or silicide gates and connections
		1, 2 or more layers of metallization or poly

components shows the need for many types of layer as summarized in Table 3.1.

All silicon technology is concerned with ways of fabricating and patterning these layers to meet extremely stringent requirements on dimensions, electrical properties, and thickness to get the best performance from the circuits. The uses of some of the layers are rather different in the two classes of IC, bipolar and MOS, and, when the electrical details are considered, they become even more distinct. The details of the fabrication processes therefore become refined for the production of one or other type of cirucit to the extent that they are usually produced in different production areas or even in different fabrication plants.

The formation of some of the layers is just a repeat of the oxide masking and doping steps used for the diode. To make the emitter region of an n–p–n transistor, for example, the surface is re-oxidized after forming the base and a second window is cut for a shallower implantation of n-type dopant over the smaller emitter region to give the doping profile shown in Fig. 3.7. It can be seen that this meets the requirements for a heavily doped emitter and lightly doped collector described in Chapter 2, Fig. 2.10(a), while the base width is the difference between the depths of the two p–n junctions.

In addition to the types of processing used for the diode there are many other steps concerned with the extra layers of an integrated circuit and we need to know something about those that have an impact on design. They are the epitaxial layer, the field oxide, and the deposited polysilicon and insulating films.

An IC factory is known in the industry as a 'fab'.

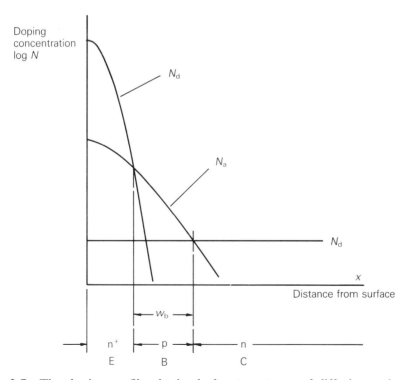

Fig. 3.7   The doping profile obtained after two stages of diffusion or ion implantation making an n–p–n bipolar transistor structure. Note that the doping concentrations are now plotted on a log scale. N_d is the donor concentration, N_a is the acceptor concentration.

Epilayer Growth

The growth of the epitaxial layer needed for bipolar ICs, and also used in some MOS processes, is achieved by depositing more silicon on to the wafer by the decomposition of gases such as silicon tetrachloride ($SiCl_4$) or silane (SiH_4) at a temperature of about 1200 °C. At this temperature the silicon atoms deposited on the surface form a crystalline layer that is continuous with the substrate. A thickness of about 3 microns is usually deposited although its effective thickness is reduced by the diffusion of the dopant from the underlying buried n^+ layer during growth.

The epilayer can be doped during growth by adding very small quantities of gases containing donor or acceptor atoms.

Field Oxidation

In both bipolar and MOS ICs, the isolation, or 'field', oxide has to be grown everywhere on the chip except in the active regions that will contain the transistors and other components. This is done by masking the active areas with a layer of silicon nitride to prevent oxidation. The nitride layer is first deposited over the entire wafer and then etched away outside the active regions using photolithography. After the growth of the field oxide, the nitride is stripped off the active regions leaving them as islands completely surrounded by field oxide. Because the field oxide is comparatively thick, at least one micron, it grows sideways under the edge of the nitride mask to give a characteristic curved edge called the 'bird's beak' as shown in Fig. 3.8. The smearing-out of the edge is one of the things that determines the minimum spacing between the active regions.

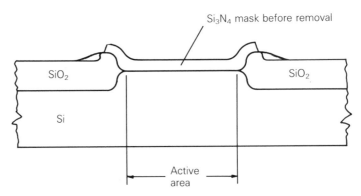

Fig. 3.8   Edge shape of the active region formed by growth of field oxide. (In some processes the silicon itself is etched before oxidation to leave a flatter surface.)

Deposited Layers

The term 'polysilicon' has been introduced on p. 30. Silicides are compounds of silicon and a metal, typically tungsten.

Integrated circuits generally need to have two or three additional layers of other inorganic materials deposited on the top surface. These include (i) a polysilicon or silicide layer to form the gates of MOS circuits, (ii) a glassy **reflow** oxide layer that, when partly melted, flows over the surface to make a smooth layer of insulation for the metallization, and (iii) a final layer of a **passivating** glass after the completion of the chip to protect it from later contamination. All these layers are deposited on the wafer by some form of chemical vapour deposition (CVD) which is a general name for the methods used to grow thin films from the

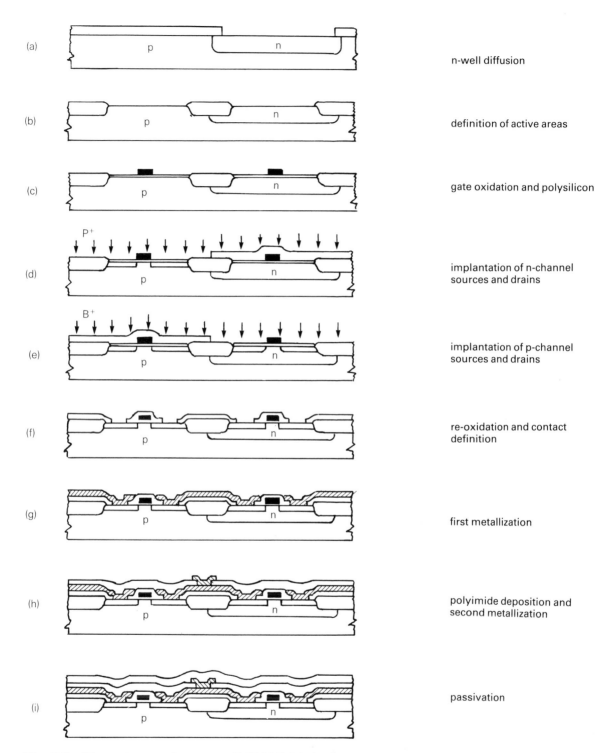

(a) n-well diffusion

(b) definition of active areas

(c) gate oxidation and polysilicon

(d) implantation of n-channel sources and drains

(e) implantation of p-channel sources and drains

(f) re-oxidation and contact definition

(g) first metallization

(h) polyimide deposition and second metallization

(i) passivation

Fig. 3.9 The main steps in an n-well CMOS fabrication process.

41

decomposition of suitable gaseous compounds. This is essentially the same as the method used for epitaxial growth but using different gases and considerably lower temperatures for the deposition of the later, non-crystalline, layers.

If two layers of interconnection are used, the insulator between them is usually a thin film of organic polyimide that is deposited after the patterning of the first layer of polysilicon or metal.

A Complete CMOS Fabrication Sequence

Figure 3.9 shows the main stages in making a very small part of an n-well, double-metal, CMOS circuit as an example of a complete fabrication sequence. For every stage shown in Fig. 3.9 there are many individual steps usually involving photolithography, with the application of photoresist, its exposure, and development, masked etching and, finally, photoresist removal. A complete process would also use additional ion implantations to control the silicon surface properties plus many stages of cleaning by washing in various chemicals.

After the initial cleaning of the silicon wafer, the process starts with the formation of the n-wells in the p-type substrate (a). As the wells are usually a few microns deep they are formed by a high-temperature diffusion of phosphorus taking several hours. In the course of this the phosphorus diffuses sideways under the edge of the masking oxide as well as downwards, and the corners of the well become rounded as shown in the cross section.

The masking oxide is etched away and the second stage is to grow the field oxide after which the nitride mask is removed from the active regions (b). The thin gate oxide is then grown by heating in an atmosphere containing very pure oxygen. This is the most critical stage in any MOS process, requiring extreme cleanliness in order to control the electrical properties of the final transistors. Polysilicon is deposited over the entire wafer and etched to leave the transistor gates (c).

The n-type source and drain of the n-channel MOST are formed by implantation of phosphorus using the hardened photoresist as a mask to cover up the places where the p-channel transistors will later be made (d). The mask used to prevent the channel of the transistor being doped is the polysilicon gate itself and this ensures that the critical gap between source and drain is exactly aligned with the gate. This is called 'self-alignment' and it is essential for making small, high-performance MOSTs. The p-channel source and drain are formed in the same way but with the implantation of boron ions (e).

The 'reflow' oxide covers the entire structure, smoothing out the edges of the surface steps and covering all the sensitive parts of the transistors. Holes are etched through it for the source and drain contacts (f), and the first layer of metal is deposited and patterned (g). The p- and n-channel transistors are shown connected together in Fig. 3.9(g) as they frequently would be in practice. The next stage is the deposition of the polyimide layer, shown with a via for the second layer of metal to connect with the common point between the transistors (h). A passivating layer of overglaze, etched away only over the contact pads for wire attachment, completes the fabrication sequence (i).

The full set of masks used by this process would be as follows:

1. n-wells,
2. Definition of active regions,
3. Polysilicon patterning,

The description here is considerably simplified. Most CMOS processes have additional implants below the field oxide and around the active regions to help reduce parasitic effects that become more important as circuit dimensions are reduced.

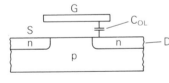

Without self-alignment, the gate layer would have to be made larger than the spacing between source and drain to be sure of covering the p-type surface even if it is displaced by the maximum manufacturing tolerance. The overlap capacitance C_{OL} that this introduces between drain and gate is a Miller feedback capacitance that greatly increases the input capacitance of the MOST when it is used as an amplifier.

Double-poly processes use two layers of polysilicon with an oxide between them.

4. n^+ doping for sources and drains of n-channel MOSTs,
5. p^+ doping for sources and drains of p-channel MOSTs,
6. Contact holes to silicon and polysilicon,
7. Metal 1 patterning,
8. Vias,
9. Metal 2 patterning, and
10. Bond-pad windows in overglaze.

Write a list of the steps needed to fabricate the bipolar transistor shown in Fig. 2.13 with its collector connected to a resistor. Include double-metal connections. How many masks are needed?

Exercise 3.1

Mask Requirements for ICs

A complete IC fabrication process usually has $10-12$ photolithography stages. Each stage uses a separate mask and the complete set of masks defines the shapes and the positions of all the doping windows, contacts and vias, and the patterning of the deposited polysilicon and metal interconnection layers. The shapes and patterns on integrated circuit masks are extremely complicated and they have to be made with great precision.

The geometrical data for making the masks is provided by the designer of a full-custom IC or by the semiconductor manufacturer in semi-custom design styles. In either case it is contained in computer files that are used for the automatic production of a set of **reticles** which are glass plates containing an enlarged image of each mask layer. The pattern is etched into a metal film on the plate by a process similar to photolithography except that it is exposed one shape at a time either optically, in a machine called an **optical pattern generator**, or by a moving electron beam in an *e*-**beam pattern generator**. In either case the machine reproduces the computer information as an exposed pattern in the photoresist which, after development, forms a mask for the etching of the metal film on the reticle plate.

The images on the reticles are further reduced in making a set of production masks with each image reproduced many times for the array of chips that will cover a complete wafer (Fig. 3.10). This is done by a **step-and-repeat camera** which projects the reduced image on to successive chip positions on the photoresist-coated mask plate. The multiple image is developed and etched into the chromium layer on the plate to form the mask for the complete array of chips. A mask of this type is mounted just above the silicon wafer in a mask aligner for projecting the pattern on to the silicon.

An alternative to using masks containing the complete array of chip images is to project a reduced reticle pattern directly on to the photoresist-coated silicon wafer in successive positions using a machine called a **direct wafer stepper**. These are used in all modern processes because they have greater resolution so that smaller shapes can be made.

The production of a set of $10-12$ reticles and masks for an IC is itself a very elaborate process. The considerable expense is justified as it can be used repeatedly for the production of many tens of thousands of chips. This explains

A direct wafer stepper is another extremely complex piece of equipment. With its control computers, clean room, and photoresist deposition and development equipment its cost is of the order of £1M.

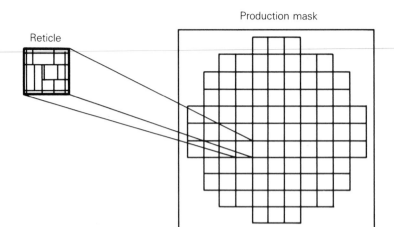

Fig. 3.10   A reticle and production mask for one layer of an IC.

why part of the semiconductor industry concentrates on very large scale production of standard ICs. For the small production runs of ASICs, however, the cost of masks may be excessive and we will find out in Chapter 6 how this can be reduced in fabricating ICs designed using semi-custom methods.

Defects, Yield and Costs

We now need to consider some aspects of the direct costs of fabricating chips once the masks have been produced. The cost is kept low by fabricating many chips simultaneously on a wafer and by processing whole batches of wafers together. The area a of a chip made by a particular fabrication process is the most important factor in determining its final cost. If the process handles wafers of area A, the number of chip areas per wafer is A/a. Since the cost of processing a wafer is the same whatever patterns it contains, the cost per chip should be proportional to its area, so that larger chips will cost more, as would be expected. The cost can be reduced, however, by using large wafers because the production costs change only slightly with A if the capital cost of the equipment can be ignored. This is why new fabrication plants are often built to handle wafers of 6″ diameter or larger.

In estimating chip costs it is necessary to allow for the defects that inevitably occur in such a complicated production process. Some defects, such as those due to processing parameters drifting out of range, affect the entire wafer, but others are localized, affecting only one circuit on the chip. Defects are often due to contamination by airborne particles, chemical residues, mask defects or crystalline faults in the silicon, and any of these can lead to defective transistors, breaks in tracks, or high-resistance contacts. All these faults can be regarded as point defects distributed almost randomly on the wafer surface.

Figure 3.11 shows a wafer with a small number of point defects which are assumed to be serious enough for any chip containing one to be rejected. The **wafer yield** Y is the percentage of chips that work correctly, leaving the remainder to be scrapped. The yield is extremely important in determining the economics of integrated circuit production. It varies greatly with chip area as can be seen by

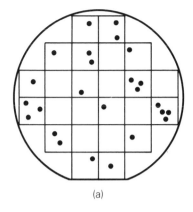 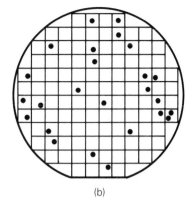

(a)   (b)

Fig. 3.11   Effect of chip area on yield. (a) and (b) have the same defects but the yields are 38% and 79%, respectively.

comparing the losses due to the same defects on wafers (a) and (b) in Fig. 3.11. In this example the yield falls from 79 to 38% when the chip edge is doubled, increasing the area by four times. Hence it can be said that the art of making ICs is in making them small enough to fit in between the defects!

If the defects are randomly distributed on the wafer the yield can be shown to fall exponentially with chip area. In practice it seems that point defects are clustered to some extent, and one better, but empirical, relationship between yield and area is

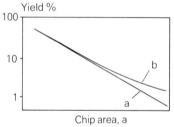

Yield against chip area, (a) random defects, (b) equation (3.1).

$$Y = 0.8/(1 + ka)^3, \tag{3.1}$$

where k depends on the total number of defects on the wafer. This expression gives a greater yield of large chips than might have been expected for the same average defect density.

It is possible to do a great deal of mathematical modelling of yield statistics, but it is of limited use because semiconductor manufacturers do not release yield information due to its commercial sensitivity. Practical manufacturing yields might be as high as 80% for small chips made by a well-established and not particularly ambitious technology, but they always fall rapidly with area and some large ICs are probably often produced at a yield of less than 10%.

The yield depends on many factors and particularly on the number of lithography stages used in the fabrication process. This explains why it is not usually desirable to add extra processing stages to make better quality resistors, for example. Yield also varies with the packing density of the circuit because smaller defects become disastrous as device sizes are reduced, although this may be counteracted to some extent by continuous improvements in the cleanliness of production areas, chemicals and gases. The yield is always lower when a new process is set up but it increases with use as sources of possible contamination are eliminated. Long-term stability of processing improves yields, which is why semiconductor manufacturers are reluctant to make changes too frequently.

Allowing for the reduction due to yield, the number of ICs per wafer falls off even more rapidly with chip size (Fig. 3.12(a)). The number of working chips produced on a wafer is $Y(A/a)$ and the cost per chip is therefore proportional to $a(1 + ka)^3$, which rises rapidly with chip size (Fig. 3.12(b)). This has important consequences for the level of integration that might be used for a particular

Obviously an equation like this can only be used over a limited range of a values. In practice it is not worth considering chips smaller than about 4 mm^2 because they are inconvenient to handle.

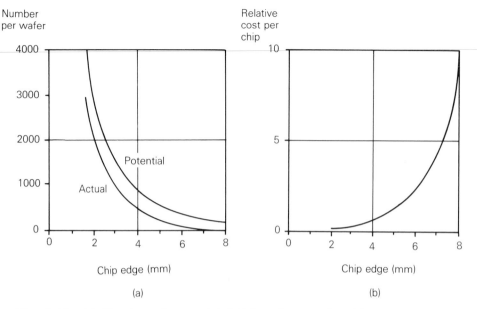

Fig. 3.12 (a) Number of potential (A/a) and actual (YA/a) chips per 6 inch wafer against chip size for $k = 0.016$. (b) Relative processing cost per chip against chip size.

application as it indicates that, beyond a certain complexity, it might be cheaper to use two small ICs than one large one.

Exercise 3.2 Calculate the relative wafer processing costs for the two chips in Fig. 3.11.
Use equation (3.1) to calculate the relative cost of a chip with an edge 1.5 times longer than chip (b) in Fig. 3.11.

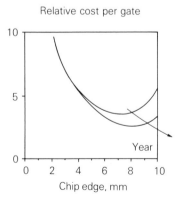

It is useful to consider the cost per gate in deciding on the best level of integration to use. Assuming that the chip area is always filled with gates at a certain density per square millimetre, the processing cost per gate is proportional to $(1 + ka)^3$ which indicates that we should always use very large numbers of the smallest possible chips to build a system at the minimum cost! However this ignores the considerable cost of packaging the chips and mounting them on PCBs. When these are included the total assembled cost per gate has a minimum value at a certain chip area. Other things being equal, this type of consideration should indicate the level of integration to be aimed at in building a system.

As the manufacturing yield improves, processing costs fall and higher levels of integration become economical. Similarly, if the gates can be made smaller without the yield falling, even more complex chips might be favoured. This explains how it is that the most advanced processes have been able to manufacture larger and larger chips each year as shown in Fig. 1.3(c).

The actual cost of making ICs depends on many factors in addition to the total production cost, and it cannot be modelled by such simple considerations as these. Every production process, however, has a limit to the maximum size of chip that can be contemplated without the yield becoming vanishingly small. The fabrication cost per gate will be lowest for small chips and it always rises rapidly with chip size towards the manufacturing limit. IC designers must understand the reasons for this and collect actual data on costs in order to decide which technology to use and whether it is better to meet a systems specification with one, two, or more ICs.

For example, a real analysis must include the cost of investment in production equipment and plant. Prices are also determined by commercial factors such as company policy and competition.

Summary

This chapter has given a very brief outline of silicon fabrication technology so that chip designers can understand how manufacturing methods influence design. All IC fabrication processes use some form of photolithography for transferring mask patterns to silicon. The patterns define the shapes of locally oxidized and doped areas, deposited conducting films, and contact holes in insulating layers. A long sequence of fabrication steps is needed to build up the structures described in Chapter 2. This is economical since large numbers of chips are made simultaneously on silicon wafers. The effect of yield on the cost of producing ICs must be appreciated for design to lead to competitive products.

The production of ICs requires extremely precise control and cleanliness at each stage. This is achieved by using very expensive automated production equipment for handling whole batches of wafers. The continuing development of devices and circuits goes hand-in-hand with improvements in fabrication methods. Wafer processing is therefore an extremely advanced subject that is the core of research and development in the semiconductor industry. Although this brief introduction is adequate for design, it is hoped that readers will be sufficiently interested to find out more about this most sophisticated technology and its underlying scientific principles. The books listed in the References are highly recommended.

The limits of fabrication technology determine the electrical performance of the circuits that will be considered in Chapters 4 and 5, and it imposes rules, the **design rules**, on the layout design which we will consider in the context of full-custom design in Chapter 9.

Problem

An IC with a chip size of (5×5) mm is being produced on 100 6-inch wafers per week with a yield of 25%. The production cost is accounted for by a wafer processing cost of £200 per wafer and a packaging cost of 75p per chip. Allowing for overheads, the production of the chip just breaks even when they are sold for £3.50 each.

The sales of this chip are limited by the number that can be produced and the demand is expected to remain constant for the year ahead. It is estimated that the IC could be redesigned as a (4.5×4.5) mm chip at a cost of £300K. Would it be wise to do this?

(It can be assumed that equation (3.1) is valid for the production process used.)

4 MOS Circuits for Digital ICs

Objectives   Studying this chapter should enable you
- ☐ To understand the basic physical limitations of all digital circuits and the criteria for comparing different circuit forms.
- ☐ To become familiar with the electrical characteristics of the MOS transistor and the gain factor β.
- ☐ To understand the operation and electrical design of CMOS and n-MOS inverter and gate circuits, and their speed of operation, including the properties and uses of MOS transistors used as switches.

To understand the properties of digital ICs we need to consider the electrical operation of the circuits that are used. Many of these contain MOS transistors, particularly in the combination of n- and p-channel types in CMOS which we will consider in greatest detail. Other circuits use bipolar transistors which have certain advantages for high-speed circuits, while the mixed form, BiCMOS, attempts to combine the best features of each.

In Chapters 4 and 5 we consider and compare the basic circuit forms shown in Fig. 4.1 which must be understood by a designer needing to select a circuit technology even for a semi-custom implementation. The chapter is also an introduction to CMOS circuits for the designer of standard cell libraries and full-custom ICs.

The basic circuits of all digital systems are inverters and gates from which all combinational and bistable circuits are built up, and we will consider how they are made in each circuit technology. In addition, single-pole switches can be made in MOS technologies and they are also used in many MOS digital circuits as will be seen.

In all digital circuit forms, the circuit diagrams do not give much scope for changes and circuit design consists of choosing a circuit type and then determining the electrical parameters to tailor it to a particular application. However, there is not a great deal that can be changed here either because the electrical parameters are fixed for the chosen fabrication process, and only the plan view dimensions can be adjusted by the circuit designer. In bipolar circuits the adjustable sizes are the areas of the transistor emitters and resistor values which can be changed over a limited range. In MOS circuits the design variables are the lengths and widths of the transistor channels, again within certain constraints. In both cases the layout design gives more scope for originality than the circuit design but it only influences the electrical performance to a small extent.

Before considering MOS circuits as such, we need to think about some of the general features of digital circuits of all types and to establish the criteria that will be used for comparing different circuit forms.

The main physical properties are the thicknesses of the layers and the doping profiles in the silicon. These determine the parameters in models of transistors, resistors, and capacitors.

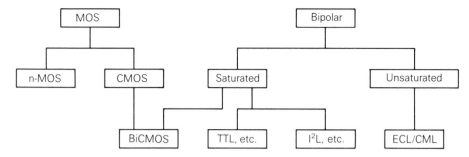

Fig. 4.1 The main circuit forms used in digital ICs.

Practical Digital Circuits

Digital Signals

Digital circuits handle information in the form of logic variables that take the state 0 or 1. The input and output voltages ideally have values V_{LO}, corresponding to logic 0, or V_{HI}, corresponding to logic 1. Where positive logic is used, V_{LO} is usually close to zero and V_{HI} is a positive voltage which may be as small as 200 mV in some circuit forms although it is usually a few volts on a PCB outside the ICs themselves. An ideal digital signal consists of a time sequence of logic 1s and 0s on one or more wires in parallel. When the signal is **synchronous** the logic state has a certain value in each half cycle of a periodic clock waveform ϕ as shown. We will generally be concerned with synchronous digital systems here because they are far easier to design than asynchronous ones due to the regular timing.

An ideal digital waveform.

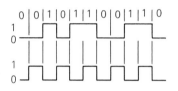

Unfortunately, practical digital signals are rather different from this ideal because of the physical limitations of circuits. The waveform of a real digital signal might be something like the little sketch here if it could be observed on an actual chip. The voltage levels are no longer distinct and the transitions are not instantaneous because all circuits take a certain time to respond to changing input voltages. The fluctuating levels are due to electrical noise, which is inherent in all circuits, and interference from adjacent circuits and the power supply. The voltages also vary between circuits due to the spread in the properties of the circuit components caused by manufacturing tolerances. All this means that logic 0 and logic 1 have to be defined by a range of voltages rather than exact values. Digital circuits therefore have to interpret the voltage levels correctly as logic 0 or 1 at certain times in order to maintain the integrity of the information.

Most synchronous ICs are required to operate at high clock frequencies to process information as rapidly as possible. If the clock frequency is f_ϕ the time per bit is $1/2f_\phi$. Typically, if f_ϕ is 20 MHz, the time is 25 ns. The transitions between logic 0 and 1 and vice versa ideally have shapes such as in Fig. 4.2 for which the time scale may be anywhere between a fraction of a nanosecond and hundreds of nanoseconds depending on the circuit form and design. The shapes of the transitions also vary with the circuit design but they can be characterized by a risetime and a falltime defined as the time for the voltage to change from 10% to 90% of the steady levels and the inverse, respectively. Other important parameters are the propagation delays $t_{p,LH}$ and $t_{p,HL}$ which are defined as the times between the input and output voltages reaching the 50% levels. The parameters for the rising edges are not necessarily the same as for the falling edges.

1 nanosecond = 10^{-9} second which is an extremely short period of time. Light travels about 30 cm in a nanosecond.

49

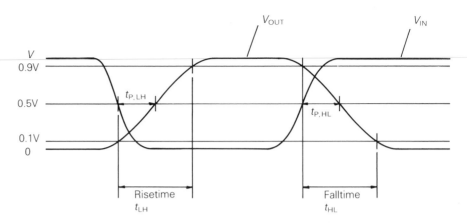

Fig. 4.2   Typical response of a digital inverter circuit with the parameters used to characterize it but omitting noise and interference.

A fanout of 4 from a NAND gate G.

One of the most important variables determining the response time of circuits is the electrical load that is driven by the output of one circuit feeding into others. The outputs of logic circuits are usually connected to the inputs of several later circuits as given by the **fan-out** F. Each input represents a certain electrical load so that the total load, and hence the response time, increases with F. Some of the circuits on an IC, such as those that distribute the clock or global control signals, may have extremely large values of F and it is necessary to design special circuits to drive these large on-chip loads as rapidly as possible. Even larger off-chip electrical loads are presented by the packaging, printed circuit board, and inputs to other ICs which have to be driven by special output or **pad-driver** circuits.

The design of digital circuits must ensure that the finite response time does not lead to errors in the logic information at the maximum clock frequency. As the clock frequency is increased the transitions get closer together, and, at the highest safe operating frequency a circuit just responds fully to one edge before the next transition comes along (Fig. 4.3). Attempts to operate the circuit faster than this will distort the digital waveform even more, eventually leading to errors.

Digital ICs may contain many thousands of circuits operating simultaneously and all taking slightly different times to respond depending on their function and design. The maximum clock frequency is determined by the slowest circuits or paths through the chip and it is important to identify such paths in the design of an IC so that their delays can be minimized. We will need to consider system timing and what happens in single clock cycles in more detail in Chapters 7 and 8. We will find that the maximum clock frequency may be more affected by logic design than by circuit design, so that, although circuits can be designed to respond in a few hundred picoseconds, the complete chip may still only operate at 10 MHz!

Circuit Performance

Many factors can be taken into account in comparing different circuit forms and designs. The three of greatest importance are speed, power, and the physical size of the circuit layout.

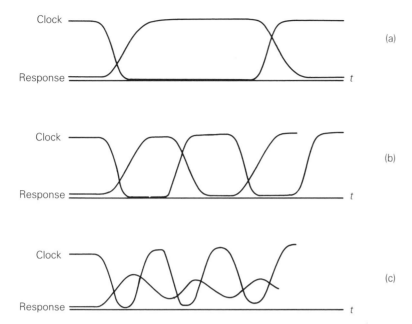

Fig. 4.3   Inverter waveforms with increasing clock frequency: (a) safe operation, (b) close to the maximum clock frequency, and (c) at a higher frequency.

1. The 'speed' of a circuit is the maximum frequency at which it can be operated reliably (Fig. 4.3). The speed needs to be high for the IC to do the maximum possible amount of digital processing work per second.
2. All circuits need a d.c. supply to provide the current for their operation. The energy drawn from the supply is dissipated in the form of heat in the circuit components. All the components in an IC are in good thermal contact with the silicon substrate which gets hot as a result. The electrical properties of silicon and hence of transistors change appreciably with temperature and ICs only function correctly up to less than 200 °C. Although the heat is conducted away quite efficiently by the package, the total power dissipation in a chip therefore has to be limited to less than about 1 watt for operation at room temperature. This can easily limit the number of circuits that can be included on the chip if they consume a lot of power. For high-density circuits the power dissipation is therefore extremely important.

   The power consumption of a circuit has **static** and **dynamic** components. The static power is due to the current taken by the circuit with fixed logic inputs, and the dynamic power is due to the extra current taken when it changes state. The average power should be evaluated for the fastest high–low, low–high switching cycle.
3. The physical size of a circuit determines the packing density on the chip if it is not limited by power dissipation, so that the smallest possible circuits are preferred. Fabrication technology determines the minimum dimensions but the size of a gate also depends on the circuit design.

In most circuit forms the speed can be increased to some extent at the expense of power and size. Two figures of merit are used to compare different circuit technologies and designs.

The maximum frequency of operation f_{max} of a gate circuit is approximately

$$f_{max} = \frac{1}{0.8(t_{LH} + t_{HL})},$$

where the rise and fall times are those of the slowest circuit path that has to respond in half a clock cycle. The path may contain bistables as well as gates.

For example, an IC will be able to contain no more than 1000 basic circuits if each of them dissipates 1 mW.

1. The first is the power-delay product $P_{AV}(t_{p.HL} + t_{p.LH})$, for a circuit with a stated fan-out F, low values being best. As the product of power and time is energy, this figure of merit is a measure of the energy used in one switching cycle and it is expressed in picojoules. Values range from about 0.01 to 50 pJ per gate for different circuit forms. Low values are best.
2. The product of the maximum number of gates per square centimetre and the maximum operating frequency, expressed in Gate Hz cm^{-2}, is a measure of the efficiency of a chip for information processing. Values are about $10^{11}-10^{12}$ for fairly complex ICs. High values are best.

These figures of merit are defined for minimum size circuits. The values for circuits containing drivers for external loads are sometimes quoted in comparing technologies and they may differ considerably from those for the internal circuits.

We will consider the different circuit technologies from these points of view at the end of Chapter 5.

The MOS Transistor

Direct-current Current Flow in the n-channel MOST

To understand MOS circuits we need to know how the transistors behave electrically. Figure 4.4 is a sketch of the structure of an n-channel enhancement MOS transistor and Fig. 4.5 shows how it is normally biased in a circuit.

The source S is taken as a ground or reference connection and positive volt-

The word 'enhancement' refers to the fact that this type of transistor is normally OFF. Its conduction is enhanced, i.e. it is switched ON when an appropriate gate voltage is applied. The term is used to distinguish this type of MOST from the depletion MOST which we will come to later.

Note that the channel length L of the MOST is equal to the width of the polysilicon track forming the gate.

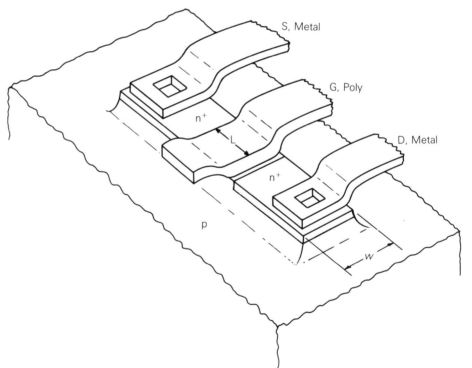

Fig. 4.4 A three-dimensional view of an oxide-isolated n-channel MOS transistor with the field oxide omitted for clarity.

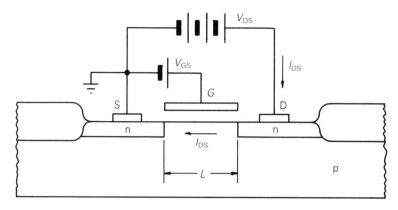

Fig. 4.5 Bias voltage directions for the n-channel MOST.

ages are applied to both the drain D and the gate G. With zero gate voltage the current between S and D is negligible because the p–n junction around the drain is reverse biased. With a positive gate voltage V_{GS} however, electrons are induced into the surface of the p-type silicon creating an n-type channel between source and drain. The induced electrons move along the channel towards the drain under the influence of the positive drain voltage V_{DS} and they are drawn through the depletion layer around the drain creating a conducting path between S and D. The flow of negatively charged electrons from S to D is equivalent to a conventional current I_{DS} from D to S.

The variation of I_{DS} with V_{DS} for various values of the controlling voltage V_{GS} is shown in Fig. 4.6(a). Current only flows for V_{GS} greater than the **threshold voltage** V_{Tn}, which is typically about 0.5 to 1.0 V. Thereafter it rises rapidly with V_{GS} as shown in Fig. 4.6(b). The characteristics of Fig. 4.6(a) can be divided into **linear** and **saturated** regions as shown. The transistor is saturated, with the drain current more or less independent of voltage, for values of V_{DS} greater than $(V_{GS} - V_{Tn})$.

It is quite straightforward to calculate the approximate current flow in an n-channel MOS transistor. In the linear region the current is

$$I_{DS} = \beta_n V_{DS}(V_{GS} - V_{Tn} - V_{DS}/2) \tag{4.1}$$

The use of the word 'linear' is rather unfortunate here. The $I_{DS} - V_{DS}$ curve is certainly linear at low V_{DS} but it bends over towards saturation. This part of the MOST characteristic is also called the **triode region**, referring to the fact that both the gate and drain voltages affect I_{DS}.

The derivation of the 'first-order' equations of the MOS transistor is given in Sparkes (1987), pp. 93–6.

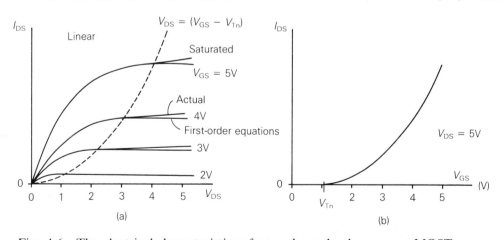

Fig. 4.6 The electrical characteristics of an n-channel enhancement MOST.

53

and in the saturated region

$$I_{DS} = I_{DS\,sat} = \frac{\beta_n}{2}(V_{GS} - V_{Tn})^2, \tag{4.2}$$

where β_n is a constant for a given device known as the **gain factor** with units of amps per volt^2. It typically has values between about 20 and $50\,\mu A\,V^{-2}$ for the small transistors used in digital ICs.

The first-order equations for the MOST do not account for the rise of current with voltage in the saturation region which is found in practice but this is of little concern as the equations are very approximate in any case. The physical operation of the MOS transistor is extremely involved, particularly when it is small, and it is still the subject of active research for sub-micron devices. Far more complicated equations are used to calculate the current in the computer simulations that will be described in Chapter 9. However, the first-order equations (4.1) and (4.2) are quite adequate for understanding the basic operation of circuits.

The gain factor β_n is important in circuit design. It is given by

$$\beta_n = C_{ox}\mu_n W_n/L_n, \tag{4.3}$$

where W_n is the width of the channel (Fig. 4.4), L_n is the length of the channel (Fig. 4.5), μ_n is the mobility of electrons in the channel (typically about $650\,cm^2\,V^{-1}\,s^{-1}$, and C_{ox} is the capacitance of the gate oxide per unit area given by the same expression as for a parallel plate capacitor

$$C_{ox} = \varepsilon_{SiO_2}/t_{ox},$$

where ε_{SiO_2} is the permittivity of SiO_2 equal to the relative permittivity, 3.9, times $\varepsilon_0 = 8.85.10^{-14}\,F\,cm^{-1}$, and t_{ox} is the thickness of the gate oxide in centimetres.

For a typical oxide thickness of 40 nm, C_{ox} is $86.3\,nF\,cm^{-2}$ which is more usefully expressed as $0.863\,fF$ per square micron.

It can be seen that all the factors in β_n are constant for a given fabrication process except for the plan-view dimensions W_n and L_n, and these only appear through their ratio. The channel length L_n is usually set to the minimum value that can be fabricated but W_n can be adjusted to some extent by the circuit designer.

For transistors with a channel length of 5 μm the difference between measured currents and the first-order model is 10–20% in the worst case. As the channel length is reduced the differences increase.

The electron mobility is a measure of how readily electrons move through the channel under the influence of the electric field due to the drain voltage. Its value, nearly always expressed in cm units in the semiconductor industry, depends on the design of the transistor.

$1\,fF = 10^{-15}\,F = 10^{-3}\,pF$

The gain factor is sometimes written as

$\beta_n = \kappa_n \cdot W_n/L_n,$

where κ_n is the gain of a square device.

Exercise 4.1 Calculate the saturation current for an n-channel MOS transistor for gate-source voltages of (a) 2 V, (b) 5 V, if $\mu_n = 650\,cm^2\,V^{-1}\,s^{-1}$, $C_{ox} = 86\,nF\,cm^{-2}$, and $V_{Tn} = 1.0\,V$.

The p-channel MOST

The p-channel MOST operates in the same way as the n-channel device but with reversed polarities (Fig. 4.7), because the current is carried by holes rather than electrons moving through a channel that, in this case, is induced on the surface of n-type silicon. The drain therefore has to be made negative with respect to the source to attract the positively charged holes and the threshold voltage V_{Tp} is also negative. Conduction only occurs when the gate is made more negative than V_{Tp}. The characteristics are as shown in Fig. 4.8. They are qualitatively the same as for the n-channel MOST but with reversed polarities and the conventional direction of the current I_{SD} is now between source and drain.

The first-order equations of the p-channel MOST are the same as for the

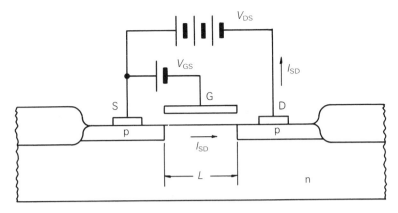

Fig. 4.7   Bias voltage directions for the p-channel MOST.

n-channel device but care has to be taken with signs. It is easiest to write the equations with | |s on the voltage symbols to indicate that the magnitude of the voltage, always a positive quantity, has to be substituted in numerically. The equations corresponding to (4.1) and (4.2) are then

$$I_{SD} = \beta_p |V_{DS}| (|V_{GS}| - |V_{Tp}| - |V_{DS}|/2) \tag{4.4}$$

in the linear region and

$$I_{SD\,sat} = \frac{\beta_p}{2} (|V_{GS}| - |V_{Tp}|)^2 \tag{4.5}$$

when the transistor is saturated, where

$$\beta_p = C_{ox} \mu_p W_p / L_p. \tag{4.6}$$

In CMOS circuits the gate oxide capacitance C_{ox} is the same for the n- and p-channel transistors because it is grown at the same time, but the hole mobility μ_p is slightly less than the electron mobility μ_n because holes do not move quite as freely through a channel as do electrons. The value of μ_p for the fabrication process giving the above value of μ_n is about $220\,cm^2\,V^{-1}\,s^{-1}$. In most CMOS processes the threshold voltage of the p-channel MOSTs is arranged to have the same numerical value as for the n-MOSTs, i.e. $|V_{Tp}| = V_{Tn}$. Hence the channel

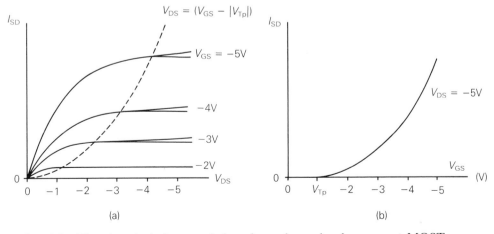

Fig. 4.8   The electrical characteristics of a p-channel enhancement MOST.

dimensions W_p and L_p are all that can be chosen by the circuit designer as for the n-channel device.

Static CMOS Circuits

MOS circuits are classed as static if their steady logic states are determined entirely by the d.c. conditions. They are the simplest and most commonly used circuits, particularly for semi-custom design, but we will also consider the alternative dynamic circuits in Chapter 10.

The CMOS Inverter

The most fundamental of all CMOS circuits is the inverter shown in Fig. 4.9. The input voltage V_{IN} may have any value between 0 and the supply voltage V_{DD}, which is typically 5 V. We will start by considering the conditions with the extreme values of V_{IN} corresponding to logic 0 and 1 respectively.

In all CMOS circuits it is useful first to identify the source and drain connections of the transistors. As the devices themselves are symmetrical, the source and drain are defined by the applied potentials. For the n-channel MOST the drain is therefore the most positive of the two, and for the p-MOST it is the most negative. Applying this rule enables us to label the circuit diagram as in Fig. 4.9(a).

The next step in thinking about CMOS circuits is to work out the gate–source voltages. Starting with $V_{IN} = 0$, V_{GS} of the n-channel MOST is zero, so that it cannot conduct and it is like an open switch. For the p-MOST, however, the source is at $+V_{DD}$ so that the gate is at $-V_{DD}$ relative to the source. As $|V_{DD}|$ must be greater than $|V_{Tp}|$ for the circuit to work, the p-channel transistor will therefore be in a conducting state almost like a closed switch and the output node will, in effect, be connected directly to V_{DD} so that $V_{OUT} = V_{DD}$.

With $V_{IN} = V_{DD}$ the situation is reversed. The n-channel MOST has a positive voltage V_{DD} on its gate so that it behaves like a closed switch while the gate and source of the p-MOST are both at V_{DD}, so that $V_{GS} = 0$, and the transistor cannot conduct. The output node is therefore connected to ground through the n-MOST and the output voltage is zero. At both extremes, the output voltage is therefore the inverse of that at the input, and the circuit produces logic inversion.

One of the main advantages of CMOS circuits is that no current is taken from the supply in either extreme so that the static power dissipation is zero. In changing state, however, the input voltage passes from 0 to V_{DD} or vice versa and both transistors can be ON at the same time allowing a current I_{DD} to flow for a short time. Analysis of this situation, Appendix 2, gives the circuit characteristics shown in Fig. 4.9(b). The transfer characteristic shows that the output changes rapidly when V_{IN} is about half way between 0 and V_{DD}. The **transition voltage** V_t is the point at which $V_{IN} = V_{OUT}$.

The inverter is said to be **electrically symmetrical** if $V_{TN} = |V_{Tp}| = V_T$ and $\beta_n = \beta_p$ which makes V_t equal to exactly $V_{DD}/2$. Substituting from (4.3) and (4.6), the βs are equal if

$$\mu_n W_n/L_n = \mu_p W_p/L_p.$$

If we furthermore make $L_n = L_p$ this gives

$$W_p/W_n = \mu_n/\mu_p \simeq 2.5 \tag{4.7}$$

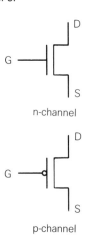

G —[D ... S

n-channel

G —o[D ... S

p-channel

We will use the symbols for the MOSTs. In this chapter, where we are only introducing CMOS circuits, we will omit the connections that have to be made to the silicon substrate and well in practical circuits. They will be added when we get into more detail on CMOS in Chapter 9.

The p- and n-channel MOSTs are examples of **pull-up** and **pull-down** devices because they connect the output to logic 1 and logic 0 respectively.

It is unfortunate that the subscript 't' is used for the transition voltage which should not be confused with the threshold voltage V_T.

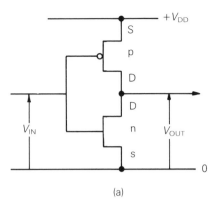

(a)

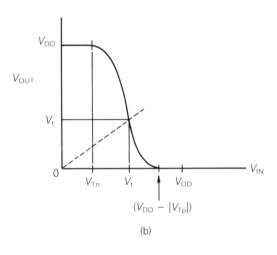

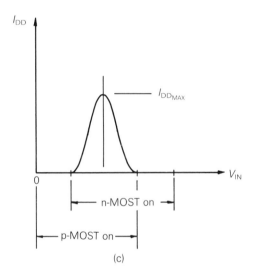

(b)

(c)

Fig. 4.9   A CMOS inverter circuit (a) and its electrical characteristics, (b) the transfer characteristic, (c) the current.

showing that the p-channel transistor should be made about 2.5 times wider than the n-MOST for electrical symmetry. In this condition the two transistors have the same conduction characteristics, the p-channel one being wider to counteract for the lower conduction due to the lower hole mobility.

The CMOS inverter has excellent d.c. characteristics. The output remains constant at V_{DD} for all input voltages between 0 and V_T, and it remains at 0 for inputs between $(V_{DD} - V_T)$ and V_{DD}, so that it rejects noise over these ranges. It also operates over a wide range of power supply voltages from about $3V_T$ upwards and it is comparatively immune to power supply noise.

CMOS Gate Circuits

The CMOS inverter circuit is easily extended to make static NAND and NOR gates (Fig. 4.10). The NAND circuit at (a) contains two n-channel **pull-down** transistors in series and two p-channel **pull-up** transistors in parallel. The only input condition in which the output is connected to ground (logic 0) is when both the n-channel transistors are ON, which is when the inputs A and B are both at

The effective mobility depends on the fabrication process and this value is typical for a p-well process. The MOSTs in a well have reduced mobility because the carriers are scattered by both donors and acceptors. Therefore with an n-well μ_n/μ_p is larger. Because of this, p-well processes are preferred for static CMOS circuits, but dynamic circuits (Chapter 10) are better in n-well.

The rejection of noise by a digital inverter circuit.

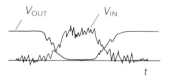

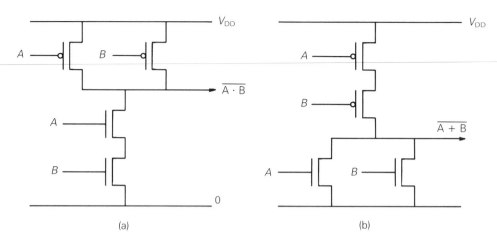

Fig. 4.10 Two-input CMOS gate circuits (a) NAND, (b) NOR.

logic 1. Also both p-channel transistors are OFF in this condition because their gate−source voltages are zero. If either A or B or both are at logic 0, at least one of the p-channel MOSTs will be ON and one of the n-channel devices will be OFF so that the output is connected to V_{DD} (logic 1). This defines the NAND function.

The operation of the NOR circuit, Fig. 4.10(b), can be deduced in exactly the same way, thinking of the transistors as switches. It is useful to remember that the n-channel transistors are switched on by logic 1 applied to the gate and the p-channel transistors are switched on by logic 0.

To get electrically symmetrical characteristics, the transistor sizes should be changed in gate circuits and this is usually done by increasing the channel width, but keeping the channel length at the minimum value. In the two-input NAND gate, for example, the width of the n-channel transistors would be doubled so that the conduction current through two in series is the same as for a single device (Fig. 4.10(a)). The same increase is made for the p-channel transistors in the NOR gate (Fig. 4.10(b)).

The number of gate inputs to CMOS NAND and NOR circuits can be increased simply by adding more series and parallel transistors to the two-input circuits. If electrical symmetry is to be maintained, the widths of the series transistors also have to be further increased. In NOR circuits the p-channel transistors, already the largest in an inverter, become extremely wide for more than about three inputs so that NAND logic is preferred. Even with NANDs it is better to avoid using gates with more than four inputs if possible because of the large silicon area used.

If one is being very precise, the increase in the width of transistors when two are connected in series should be nearer 3 times to maintain the same speed. This is because the transient voltage is not divided equally between the transistors and their total resistance is increased.

Circuit Speed and Power

As already explained, the speed at which the output of an inverter or gate circuit changes determines the operating frequency of digital systems. In general the speed is determined by the redistribution of charge both inside the transistors and in the circuit capacitances. For the MOS transistor, the channel forms extremely rapidly when the gate voltage is first applied so that the speed of response is determined entirely by the speed of charging and discharging the circuit and device capacitances.

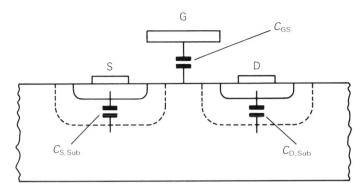

Fig. 4.11   The main capacitances associated with an MOS transistor.

The most important capacitances associated with the MOS transistor are shown in Fig. 4.11 for an n-channel device. The gate capacitance C_{GS} is the parallel-plate capacitance between the gate and the channel. With channel dimensions of W_n and L_n, C_G is just $C_{ox}W_nL_n$, where C_{ox} is the gate capacitance per unit area. The other two capacitances $C_{S,sub}$ and $C_{D,sub}$ are the depletion layer capacitances of the source and drain p–n junctions to the substrate or well in which the transistor is made. These capacitances change with the reverse bias on the junctions (Appendix 1). In an inverter the sources of both transistors are at fixed potentials so that the source–substrate capacitances are unimportant.

All the capacitances at the output node of a CMOS inverter can be lumped together into a single load capacitance C_L, as shown in Fig. 4.12. C_L has the following main components.

1. The input capacitance C_{IN} of the following stage which is given by the sum of its gate capacitances, i.e.

$$C_{IN} = C_{ox}(W_pL_p + W_nL_n) \qquad (4.8)$$

   If the output has a fan-out to F identical gates the total input capacitance will be FC_{IN} and this is the largest component of C_L in most cases.
2. The **wiring capacitance** which is associated with all the connections between the drains of both output transistors and the gates of the following stage. These interconnections are made of metal or polysilicon tracks insulated by oxide films from the silicon.
3. The depletion layer capacitances of the p–n junctions forming the drains of the p- and n-channel transistors of the driving inverter.

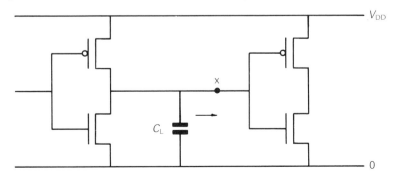

Fig. 4.12   The load capacitance C_L at the output node of an inverter.

When the inverter input changes from logic 1 to logic 0, C_L is charged through the p-channel MOST and the n-MOST is turned off (Fig. 4.13(a)). For a ramp input the time for the output to rise to $V_{DD}/2$ is

$$t_{pLH} = 2.57C_L/(\beta_p V_{DD}) \tag{4.9}$$

assuming that the input and output risetimes are the same (Appendix 3).

For the opposite transition, with the input changing from 0 to logic 1, C_L discharges through the n-channel MOST (Fig. 4.13(b)) and the delay time is given by a similar expression,

$$t_{pLH} = 2.57C_L/(\beta_n V_{DD}). \tag{4.10}$$

The rising- and falling-edge delays are equal if the βs are the same, which is the advantage of electrical symmetry. In practice, however, the transistor dimensions are sometimes made the same for both devices. The rising-edge delay is then between 2 and 3 times greater than for the falling-edge because of the greater resistance of the p-channel device.

Exercise 4.2 Calculate the rising- and falling-edge delays of a CMOS inverter with a fan-out of 4, given that $\mu_n = 650\,\text{cm}^2\,\text{V}^{-1}\,\text{s}^{-1}$, $\mu_p = 240\,\text{cm}^2\,\text{V}^{-1}\,\text{s}^{-1}$, $C_{ox} = 0.86\,\text{fF}\,\mu\text{m}^{-2}$, $L_n = L_p = 2.5\,\mu\text{m}$, $W_n = W_p = 4.0\,\mu\text{m}$, and $V_{DD} = 5\,\text{V}$. Assume that the total wiring and drain capacitance is 40 fF.

From (4.7) it might be thought that a CMOS circuit could be made to operate faster by increasing the β values by making the transistors wider. The problem in doing this is that the gate capacitances are also increased so that the load on the previous stage is increased and there is unlikely to be much overall benefit. For an electrically symmetrical inverter the delay t_p can be expressed in terms of the input capacitance C_{IN} as

If the wiring and device capacitances are ignored C_L/C_{IN} is approximately equal to the fan-out of a gate.

$$t_p = \tau \frac{C_L}{C_{IN}}, \tag{4.11}$$

where τ is a characteristic of the technology given by

Equation (4.12) is easily derived by substituting $\beta_n = C_{ox}\mu_n W_n/L_n$ into (4.10) and expressing C_{ox} in terms of C_{IN} from (4.8).

$$\tau = \left(\frac{2.57L^2}{\mu_n V_{DD}}\right)\left(\frac{W_n + W_p}{W_n}\right). \tag{4.12}$$

Taking $L = 2.5\,\mu\text{m}$, $\mu_n = 650\,\text{cm}^2\,\text{V}^{-1}\,\text{s}^{-1}$, $V_{DD} = 5\,\text{V}$, and $W_p = 2.5W_n$, for example we find that $\tau = 0.17\,\text{ns}$. In practice C_L/C_{IN} is at least 2 and it is often far greater, but the calculated delays are still no more than about 1 ns. Similar reasoning can be applied to gate circuits.

It must be emphasized again that the first-order equations for the MOSFET become less accurate as L is reduced, and delay values based on these equations give estimates of circuit speed that need to be checked by simulation if they are critical. Circuit simulation using SPICE is described in Chapter 9.

For $L = 3\,\mu\text{m}$ and $C_L = 166\,\text{fF}$; equation (4.12) gives a delay of 0.80 ns. The simulation result in Fig. 9.7 gives 1.0 ns.

The current taken from the supply when a CMOS circuit changes state has two components. The first is the current that flows directly from V_{DD} to ground through both the transistors of an inverter for example (Fig. 4.9(c)). The second is the additional current that flows into C_L when it is charged. If the transitions

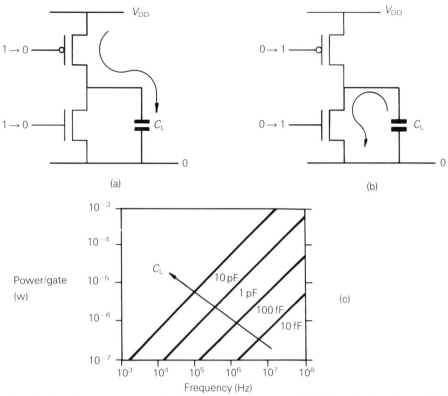

Fig. 4.13 The current paths in (a) the charging and (b) the discharging of C_L. The resulting power dissipation per gate is shown at (c) for various values of C_L and $V_{DD} = 5\,\text{V}$.

are fairly rapid the charging current is by far the largest and we can easily calculate the resulting dynamic power by considering the energy transfer. The energy dissipated in one cycle of charging and discharging a capacitor C is simply CV^2. In this case, if C_L is charged to V_{DD}, f times per second, the power P is

$$P = C_L V_{DD}^2 f. \tag{4.13}$$

This accounts for nearly all the power consumption of a CMOS circuit. Figure 4.13(c) shows how the power dissipation per gate varies with frequency for various values of C_L as calculated from (4.13). Some simple calculations will show that this has important implications for the future of highly complex CMOS circuits operating at very high frequencies even though many of the gates on an IC may not change state at the clock frequency.

With a fan-out of 3, CMOS gates typically have C_L about 110 fF for a 2.5 μm process and about 25 fF for a 1 μm process.

Calculate (a) the power, and (b) the power–delay product of the CMOS inverter in Exercise 4.2 operating at its maximum frequency which is approximately $1/4t_p$ Hz.

Exercise 4.3

Calculate the power dissipation of 10 000 of the above circuits integrated on a single IC.

Exercise 4.4

---

n-MOS Circuits

The earliest MOS ICs used only p-channel transistors because they are easier to make. However n-channel MOSTs have better performance due to the higher mobility of electrons compared with holes in a channel, and p-MOS processes are now obsolete.

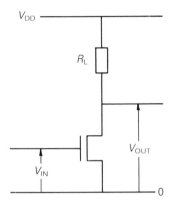

A simple n-MOS inverter.

The load-line construction is a graphical way of solving Kirchoff's voltage law. Here the load is very non-linear but the method is used in the same way as for a linear resistive load.

As the name implies, n-MOS circuits contain only n-channel transistors. Although rather overshadowed by the more recent CMOS circuits, n-MOS still has certain advantages and dynamic circuit forms have a performance which is very comparable to that of CMOS.

A static n-MOS inverter uses a single transistor with a resistive load R_L. The load needs to be of fairly high value, about $20\,k\Omega$, but it does not need to be particularly linear, and in an IC it is most conveniently made from another MOS transistor. Early n-MOS circuits used a normal n-channel transistor with the gate connected to the drain as the load resistor. However, this has long been replaced in digital MOS circuits by a far better type of resistor made using a modified type of transistor known as a **depletion MOST**.

The depletion MOST has the same structure as a normal, or enhancement, n-channel MOST, but it has a permanently conducting channel between source and drain formed by an extra, very thin, n-type layer in the silicon surface. This permanent channel is indicated by the thickened bar on the symbol for the device (Fig. 4.14). Electrically, the depletion MOST has very similar characteristics to the enhancement device but it has a negative threshold voltage V_{TD} so that it conducts even with $V_{GS} = 0$ unlike the enhancement transistor. Its main application is as a resistor which is made by connecting the gate to the source, so that $V_{GS} = 0$, giving the two-terminal characteristic shown in Fig. 4.14. The saturated current $I_{DS,\,sat}$, is given by the same expression as for an enhancement transistor (equation (4.2)), but with the changed threshold value V_{TD} and $V_{GS} = 0$, i.e.

$$I_{DS,\,sat} = \beta_D |V_{TD}|^2/2, \tag{4.14}$$

where β_D is the gain factor of the depletion MOST.

The use of a depletion resistor in the n-MOS inverter is shown at (a) in Fig. 4.15. The circuit is analysed by the load-line construction at (b), where the characteristic of the load resistance is drawn 'backwards' starting at V_{DD} on the voltage axis in the normal way. For a particular input voltage V_I the intersection of the $I-V$ curves for the pull-down and load devices at P determines the output voltage V_O. The construction is used to plot the transfer characteristic of the inverter at (c).

With V_{IN} at 0 the pull-down transistor is OFF and the output is clamped to the full logic 1 voltage of V_{DD}. However, with a logic 1 input both devices are

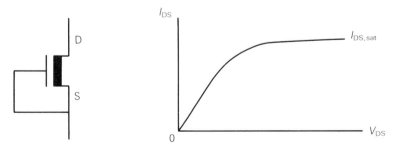

Fig. 4.14   The $I-V$ characteristic of an n-channel depletion transistor connected as a resistor.

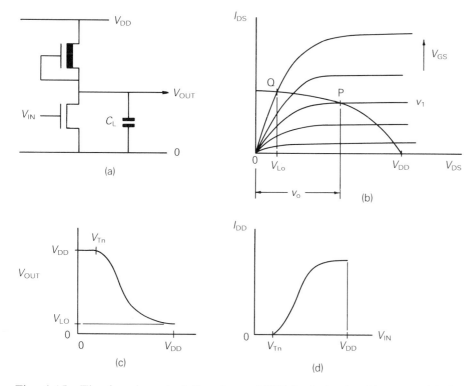

Fig. 4.15   The d.c. characteristics of an n-MOS depletion load inverter, (a) the circuit, (b) the load-line construction, (c) the transfer characteristic, and (d) the current characteristic.

conducting and the output voltage V_{LO}, corresponding to logic 0, is not zero because the circuit is a potential divider in this condition. V_{OUT} is therefore a proportion of V_{DD} determined by the ratio of the effective resistances of the two devices. In designing the circuit, V_{LO} has to be made low enough to transfer the logic state reliably to later stages. A somewhat arbitrary choice of V_{LO} would be to make it no more than $V_{Tn}/2$. We can find out how to achieve this value by equating the currents through the two transistors at the operating point Q in Fig. 4.15(b). The pull-down transistor is in the linear region at this point and its current is given by equation (4.1) with $V_{DS} = V_{LO}$ and $V_{GS} = V_{DD}$. The load transistor is saturated so that its current is given by (4.14). Equating the two currents

$$I_{DS\ pull\ down} = I_{DS,\ sat\ load}$$

or $$\beta_n V_{LO}(V_{DD} - V_{Tn} - V_{LO}/2) = \beta_D |V_{TD}|^2/2. \qquad (4.15)$$

Equation (4.15) is used to find β_n/β_p in the design of the inverter. This is most easily illustrated by a numerical example.

Determine the channel lengths for the transistors in an n-MOS inverter made using a 3 μm process if the channel widths are to be equal. The voltage values for the fabrication process are $V_{DD} = 5\,V$, $V_{Tn} = 1.0\,V$, $V_{TD} = -4.0\,V$, $V_{LO} = 0.5\,V$. *Solution*   By substituting the voltage values into (4.13) we find

Worked example 4.1

$\beta_n/\beta_D = 4.27$.

For each device

$$\beta = \mu C_{ox} W/L$$

and C_{ox} is the same for both so that

$$\frac{\mu_n}{\mu_D} \frac{W_n}{W_D} \frac{L_D}{L_n} = 4.27. \qquad (4.16)$$

The mobility μ_D in the channel of the depletion MOST is slightly less than in the enhancement device because of the greater scattering of the electrons by the extra dopant atoms. Taking μ_n/μ_D as 1.2, we find that for equal channel widths

$$\frac{L_D}{L_n} = 3.56.$$

In practice, channel dimensions must usually be expressed in multiples of 1.0 or 0.5 μm. The designer might therefore use $L_n = 2.5$ μm and $L_D = 9.0$ μm for this inverter to be certain to get the required value of V_{LO}.

The speed of an n-MOS inverter is determined by the charging and discharging of the load capacitance C_L (Fig. 4.15(a)) in the same way as for CMOS. A most important difference, however, is that it takes longer to charge C_L than to discharge it in n-MOS because the β value of the load is greater than for the pull-down transistor (by the factor 4.27 in the example above). This lack of symmetry can lead to problems in the timing of n-MOS circuits.

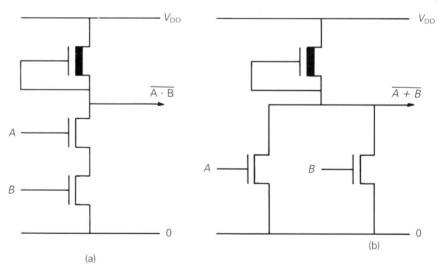

(a)

(b)

Fig. 4.16   n-MOS gate circuits (a) a two-input NAND gate, (b) a two-input NOR gate.

Static n-MOS gate circuits are made by adding pull-down transistors to the basic inverter circuit as shown in Fig. 4.16. In the NOR gate each pull-down can be the same size as in the inverter so that the worst-case value of V_{LO} is unchanged. For a two-input NAND gate, however, with two transistors in series, their widths have to be doubled to give the same total current flow when both are ON together. With a larger number of inputs the pull-down transistors in a NAND gate become excessively large so that NOR logic is preferred for n-MOS ICs.

The main problem inherent in static n-MOS circuits is that d.c. current flows continuously with the output at logic 0 as shown in Fig. 4.15(d). This should be compared with Fig. 4.9(c) for the CMOS case. The static power dissipation per gate is $I_{DS\,max}V_{DD}/2$ assuming that a circuit has a logic 0 output for half the time on average, and there is additional dynamic power dissipated on changing states as for CMOS circuits. The total power dissipation is therefore greater than for CMOS and this can limit the number of n-MOS circuits that can be put on to a single IC.

This problem is partly overcome by using dynamic circuit forms that effectively isolate circuits from the supply for part of every clock cycle, leaving the logic state determined by the charge stored on the gate capacitances. There are many different ways in which this can be done and some of them are used extensively in high-density standard-product ICs. As device dimensions decrease, however, the power consumption becomes an increasing problem and only CMOS can be used for the highest density chips.

In spite of the problem with power, n-MOS is still attractive for high-speed ICs partly because gate circuits are simpler since only one transistor has to be switched per logic input compared with the two (one n-channel and one p-channel) in CMOS. The gate input capacitance (equivalent to equation 4.8) is therefore reduced by a factor of at least 2, which increases the speed, and the layout can also be more compact. The fabrication of n-MOS requires an extra ion implantation stage to form the channels of the depletion MOSTs and additional stages to make the contacts between the polysilicon gates and sources of these devices, which never occurs in CMOS. On the other hand the stages involved in fabricating the wells for the p-channel MOSTs in CMOS are no longer required.

Most n-MOS circuits use **substrate bias** to reduce the source and drain capacitances and increase the speed. The circuit ground is made positive with respect to the substrate, typically by about 2.5 V. This voltage is often generated on the chip.

MOS Switches and Bistables

We have seen that enhancement MOS transistors can be regarded as single-pole switches, being open or closed depending on the gate–source voltage. The switching properties are used extensively in many other circuits and particularly in MOS bistables. It is necessary to understand how MOSTs work electrically as switches to design such circuits.

The MOS Pass Transistor

We start by considering the use of an n-channel MOST as a switch connecting a positive voltage source V_{IN} ($= V_{DD}$) to a capacitive load C_L, as in Fig. 4.17(a). This circuit has no permanent connections to a d.c. supply but the voltages are still referred to the circuit ground potential, zero. A MOSFET used in this way is called a **pass transistor**.

We will assume that C_L is initially discharged and that a voltage V_G, equal to the full logic 1 voltage V_{DD}, is applied to the gate at $t = 0$. We must first identify the source and drain of the MOST, the drain being the most positive of the two. For $t > 0$, current I_{DS} will flow through the transistor, charging C_L so that V_{OUT} rises as shown at (b) in Fig. 4.17. As C_L charges, V_{GS} falls and when it reaches the threshold voltage V_{Tn} the MOST turns off and the current ceases. This

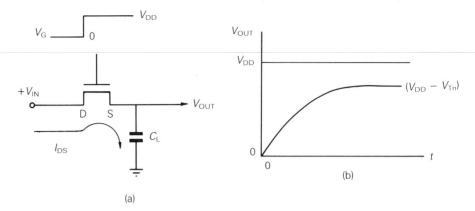

(a)

(b)

Fig. 4.17   The charging of a capacitance through an n-channel MOST.

occurs when V_{OUT} reaches $(V_{DD} - V_{Tn})$ and this is the maximum voltage that is transmitted by the switch. The output is said to have a **threshold voltage drop** relative to the gate voltage and this has important consequences for the transmission of digital signals through pass transistors. Circuits connected to pass transistors therefore have to be specially designed to accept the reduced logic 1 voltage level.

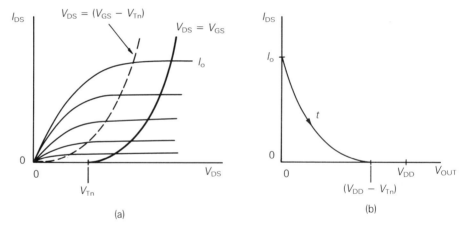

(a)

(b)

Fig. 4.18   (a) The locus of points with $V_{DS} = V_{GS}$ on the characteristic of an MOS transistor, and (b) the current flow in charging C_L in the circuit of Fig. 4.17.

This treatment of the MOST as a switch is somewhat simplified because we are using the first-order MOST equations throughout this book. A more detailed consideration of the operation of the MOST shows that the effective threshold voltage varies with the source-substrate voltage. This is known as the **body effect**. It does not substantially alter the description here.

To take this a little further, we need to know how the current varies during the charging of C_L. The voltages V_G and V_{IN} applied to the transistor are both equal to V_{DD} at $t = 0$, and the current I_0 is just the saturation current with V_{GS} equal to V_{DD}. As the source voltage rises, V_{DS} remains equal to V_{GS} and we need to identify this condition on the transistor characteristics (Fig. 4.18(a)). Noting that saturation occurs at points where $V_{DS} = (V_{GS} - V_{Tn})$, the locus of $V_{DS} = V_{GS}$ must be a parallel curve displaced by an amount V_{Tn} along the voltage axis. The output voltage is $(V_{DD} - V_{DS})$ so that the current curve can be transferred to a graph against V_{OUT} as shown at (b). This confirms that conduction ceases when the output reaches $(V_{DD} - V_{Tn})$.

The situation on discharging a capacitor through an n-channel pass transistor is

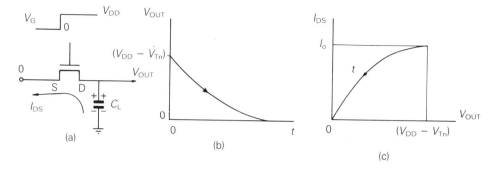

Fig. 4.19 The discharging of a capacitance through an n-channel MOST.

shown in Fig. 4.19. We assume that C_L has previously been charged to $(V_{DD} - V_{Tn})$ and that the switch input is now grounded. The gate voltage $V_G = V_{DD}$ is applied at $t = 0$. With these polarities the source and drain are reversed and V_{GS} remains constant at V_{DD} while C_L discharges, so that V_{OUT} eventually falls to zero (Fig. 4.19(b)) and there is no voltage drop in the discharge direction. The current locus in (c), which is the transistor characteristic curve for $V_{GS} = V_{DD}$, confirms this.

The above reasoning can also be applied to the use of a p-channel MOST as a pass transistor and it is left to the reader to show that the problem of voltage transfer is reversed in this case. With a p-channel pass transistor the output charges to the full value of the input voltage but it discharges to only $|V_{Tp}|$ before ceasing to conduct, leaving a residual voltage on C_L. The ways in which the current varies with output voltage during charging and discharging are also the reverse of those with the n-channel transistor.

The current loci on charging and discharging C_L through a p-channel MOS pass transistor.

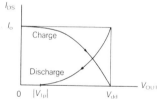

The CMOS Transmission Gate

The threshold voltage drops in the charging direction with an n-channel pass transistor, and in the discharging direction with a p-channel device, must be allowed for in the design of subsequent circuits if either is used alone. The **transmission gate** overcomes the problem in CMOS by using n- and p-channel

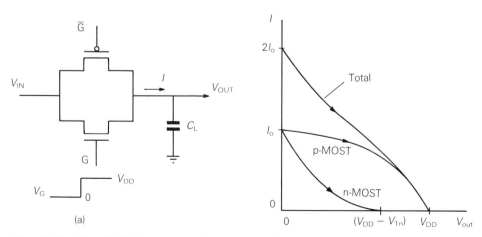

Fig. 4.20 The CMOS transmission gate and (b) the current flow through the two transistors on charging a capacitor.

transistors in parallel to make an almost ideal bi-directional switch (Fig. 4.20). The transistor gates are driven by logic signals G and $\bar{G}$ that are the inverse of each other so that both transistors are turned on together. In charging a load capacitance, current flows initially through both transistors in parallel. When the output voltage reaches $(V_{DD} - V_{Tn})$ the n-channel MOST ceases to conduct but current still flows through the p-channel device until the output is fully charged to V_{DD}. In the discharge direction, it is the p-channel MOST that stops conducting first, leaving the n-channel device to carry the current until the output voltage falls to zero. The deficiency of each transistor is therefore compensated for by the other.

If the β and V_T values of the transistors are the same, the initial charging current is $2I_0$, where I_0 is the saturation current in either transistor with $V_{GS} = V_{DD}$. Although the currents subsequently vary with output voltage in different ways (Fig. 4.20(b)) their sum varies almost linearly with voltage so that the transmission gate has an almost ohmic resistance R_{TG}, approximately equal to $V_{dd}/2I_0$. Using the expression for I_0 for the n-channel MOST

$$I_0 = \frac{\beta_n}{2}(V_{DD} - V_{Tn})^2$$

gives

$$R_{TG} = \frac{V_{DD}}{\beta_n(V_{DD} - V_{Tn})^2}. \tag{4.17}$$

Charging and discharging a capacitance C_L through a transmission gate follows the normal exponential law with a time constant $R_{TG}C_L$. Delays through transmission gates are comparable to those through inverters.

Worked example 4.2   Calculate the time constant for charging a capacitance of 100 fF to 5 V through a transmission gate in which each transistor has $\beta = 40\,\mu A\,V^{-2}$ and $V_T = 1\,V$. *Solution*   Substituting these values into (4.17), $R_{TG} = 4.9\,k\Omega$. Therefore $T = R_{TG}C_L = 0.5\,ns$.

Transmission gate symbols.

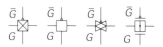

There are many ways of making D-type bistables using gates, one of which is

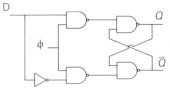

This circuit could be built using any circuit technology but it is often not the best way of making a D-type.

The transmission gate is an extremely useful CMOS circuit element for which various circuit symbols are used.

MOS Bistable Circuits

Bistable circuits, also called **flip-flops**, are needed in all digital systems for the temporary storage of data. Although they can be built up from gates, simpler circuits can be made in MOS technologies using switches.

The basic static storage element in MOS ICs is a D-type bistable consisting of two inverters with positive feedback as in Fig. 4.21(a). This circuit has stable states with either 0 or 1 at the output. To use it as a storage element we need to control the state and that is done by disconnecting the feedback and simultaneously connecting the data input D to the inverters (Fig. 4.21(b)). The switches are normally controlled by a clock ϕ, and if S_1 is closed when $\phi = 1$, S_2 must be closed when $\phi = 0$, which is equivalent to S_2 being controlled by $\bar{\phi}$.

In n-MOS bistable circuits the switches used are single pass transistors but in CMOS they are transmission gates (Fig. 4.21(c)). This circuit is a level-sensitive, or **transparent**, D-type latch with the symbol as at (d). It is the basic bistable

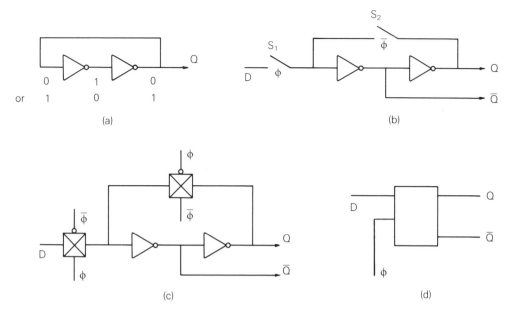

Fig. 4.21 (a) A feedback data storage circuit, (b) a practical implementation as a transparent D-type, (c) a CMOS D-type latch, and (d) its circuit symbol. Circuits of this type, controlled by a single clock ϕ, require very careful design.

element used in MOS ICs and other types, such as edge-triggered D-types, T-types, etc., are built up from combinations of the transparent D-type and various gates. The most important of these are described in Appendix 4 which we will need to reconsider in the context of digital design in Chapter 7.

Although bistable circuits such as these are commonly used in cell libraries, they need to be treated with caution in full-custom design because of the possibility of incorrect operation if the edges of the ϕ and $\bar{\phi}$ signals are not exactly coincident. This problem, caused by **clock skew**, is overcome by using a two-phase clocking scheme that will be described in Chapter 7.

Summary

This chapter has described the basic static CMOS and n-MOS circuits that are commonly used in digital ICs. The operation of these circuits is determined by the I–V characteristics of the transistors, as given approximately by simple first-order equations, (4.1) and (4.2). The current is proportional to the transistor gain factor β, the value of which depends on the channel dimensions. CMOS and n-MOS inverters and gates with good transfer characteristics can be designed using these equations.

All circuits introduce a delay between the input and output transitions. In MOS circuits the delay depends on the load capacitance and hence on the fan-out of the gate. The delay is easily estimated for the design of high-speed digital ICs. Every time a MOS gate changes state, electrical energy is converted into heat and the total power dissipation can be a problem for the design of very high-speed VLSI ICs. The problem is far greater for n-MOS due to the additional static power dissipation in this circuit form.

Digital MOS ICs commonly use transistors as switches. A single MOS transistor is unable to transmit the full range of voltages, 0 to V_{DD}, that are applied to the gate. This problem is overcome in CMOS by using one transistor of each polarity in the **transmission gate**. Switches are used to advantage in the design of MOS bistable circuits.

Problems

All the Problems below refer to a 2 micron CMOS fabrication process with the following parameters:

Gate oxide thickness	35 nm
Electron mobility in n-channel	$580\,\text{cm}^2\,\text{V}^{-1}\,\text{s}^{-1}$
Hole mobility in p-channel	$200\,\text{cm}^2\,\text{V}^{-1}\,\text{s}^{-1}$
Threshold voltages	0.8 V
Minimum transistor width	2.5 μm
(Dimensions to be multiples of 0.5 μm)	
Supply voltage, V_{DD}	5.0 V

(Permittivity of gate oxide $= 3.45 \times 10^{-13}\,\text{F cm}^{-1}$)

4.1 What is the gate capacitance of a minimum size n-channel MOST and its gain factor β?

4.2 What is the size of a p-channel MOST to have the same gain factor as for the n-channel device?

4.3 What is the saturation current of the transistor in (4.1) for gate–source voltages between 1 and 5 V in steps of 1 V?

4.4 Find the transition voltage of an inverter made with p- and n-channel transistors of the same dimensions. (Hint: Both transistors are saturated at V_t.)

4.5 What is the value of τ for an electrically symmetrical inverter made by this process?

4.6 The inverter in (4.5) is connected to four 3-input NAND gates on the same chip. Calculate the resulting propagation delay.

4.7 What is the delay in (4.6) if the interconnections have an additional capacitance of 60 fF?

4.8 What is the power dissipation of the inverter in (4.7) if it operates at 20 MHz?

4.9 What is the resistance of a minimum size transmission gate? What is the delay time for charging the load in (4.7) through a transmission gate connected to V_{DD}?

Bipolar Circuits for Digital ICs   5

Studying this chapter should enable you

□ To consolidate your understanding of the electrical characteristics of the bipolar transistor and its use as a switch.

□ To understand the difference between saturated and non-saturated switching of the bipolar transistor.

□ To distinguish between the most commonly used bipolar digital circuits forms, TTL, STL, ECL and others, in order to appreciate the operation of bipolar cell library circuits used in semi-custom IC design.

□ To compare the performance of all types of bipolar and MOS digital circuit forms for use in ICs.

□ To appreciate the advantages of BiCMOS circuits containing both MOS and bipolar transistors.

Objectives

The Bipolar Transistor

Current Flow in the n–p–n Bipolar Transistor

Bipolar integrated circuits use n–p–n transistors, the structure of which have been described in Chapter 3. As in the MOS transistor, the current through the device is controlled by the voltage applied to the input. In the bipolar transistor the current flow is between the collector and emitter connections which correspond to the drain and source of the MOST and the controlling voltage is applied between base and emitter, corresponding to the gate–source voltage of the MOST. Although the electrical characteristics of the two classes of transistor are somewhat similar, the internal operation is quite different and this has important consequences for the circuits in which they are used.

The control action in the bipolar transistor is produced by the interaction of the two p–n junctions in the n–p–n structure. We saw in Chapter 2 that the p–n junction is an excellent electrical rectifier with a characteristic such as Fig. 2.1. A first-order expression for the d.c. current through a p–n junction with an applied voltage V is

In the reverse direction the p-type is made negative. In the forward direction the p-type is made positive.

$$I = I_s[\exp(qV/kT) - 1], \tag{5.1}$$

where I_s is a parameter known as the saturation current and kT/q is the **thermal voltage** which has the value of 25.9 mV at an absolute temperature T of 300 K, which is about room temperature.

In the reverse direction V is negative in (5.1) so that the exponential term $\exp(-qV/kT)$ becomes negligible and the reverse current I_r is equal to $-I_s$ and independent of voltage above about 100 mV.

In the forward direction the voltage is V_f and positive. For $V_f \gg kT/q$ the '−1' in (5.1) becomes negligible compared with the exponential term and

q is the electronic charge, 1.6×10^{-19} C. k is Boltzmann's constant, 1.38×10^{-23} J K^{-1} It is not necessary to remember these but kT/q is a very important parameter throughout electronics.

I_r does eventually increase with voltage but beyond the voltage levels at which it is used in a circuit.

$$I_f = I_s \exp(qV_f/kT), \tag{5.2}$$

71

i.e. the forward current rises exponentially with voltage.

Exercise 5.1 Calculate the forward current through a p–n junction having I_s equal to 10^{-14} A for voltages rising in 100 mV steps up to 700 mV.

Exercise 5.2 What is the difference between the forward voltages of two p–n junctions carrying the same current if one of them has a saturation current of 10^{-12} A and the other 10^{-13} A?

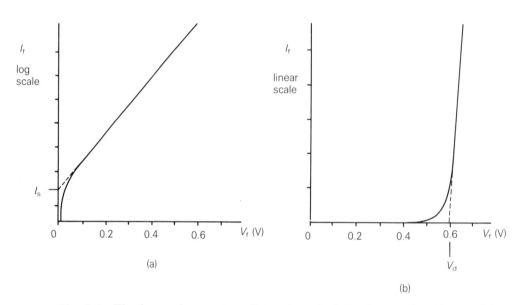

(a)

(b)

Fig. 5.1   The forward current–voltage characteristic of a p–n junction on (a) a logarithmic and (b) linear scale of current. In (a) the current increases by 10 times for every 59.6 mV increment of V_f.

The saturation current of p–n junctions in ICs is very small, 1 pA or lower. Although the forward current rises continuously with voltage on a logarithmic scale (Fig. 5.1(a)), it remains small in circuit terms up to about 500 mV. On a linear scale, however, the current rises rapidly beyond this (Fig. 5.1(b)) and in circuits one can often consider the diode to be 'switched on' for applied voltages greater than the diode voltage drop V_d, even though there is no actual switching action in the junction. V_d is often taken to be about 0.6 V. The current through a forward-biased diode has to be limited by the circuit to prevent it from being destroyed by any voltage in excess of about 1.1 times V_d.

In general, the forward current in a p–n junction is carried by the flow of both electrons and holes. Carriers of both types move through the depletion layer because the potential barrier between the p- and n-type regions is reduced by the applied forward voltage V_f. However, if the junction is made with the n-type silicon far more heavily doped than the p-type, forming an n$^+$p junction, the forward current is carried predominantly by electrons that are **injected** from the n$^+$ into the p-region by passing through the depletion layer. The injected electrons

The current flow in p–n junctions and transistors is described in detail in Sparkes (1987).

move through the p-type layer to a metallic contact which completes the circuit for current flow.

In the n–p–n transistor the emitter region is heavily doped n^+ in this way so that the current in the forward-biased emitter–base junction is almost entirely due to the injection of electrons into the p-type base. The transistor structure places the collector junction just below the emitter and the injected electrons are captured or **collected** by the reverse-biased base–collector junction, increasing the collector current at the expense of base current (Fig. 5.2). The flow of the

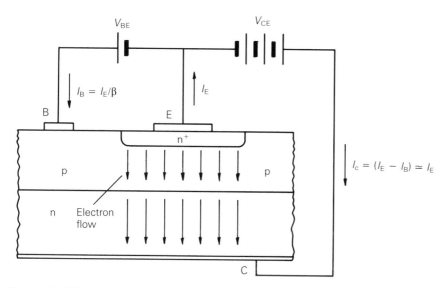

Fig. 5.2   The current flow in a normally biased n–p–n transistor structure.

injected electrons to the collector rather than the base terminal is called **transistor action**. It produces an electron current flowing directly between emitter and collector but under the control of the voltage between the emitter and the base.

If transistor action was perfect, the base current would be zero and I_C and I_E would be exactly equal. However, the flow of electrons to the collector can never be 100% efficient so that there is always a small base current I_B equal to I_C/β, where β is the **common-emitter current gain** of the device, typically between about 50 and 200 or even more. The relationships between the resulting currents are shown in Fig. 5.2. Note that the flow of negatively charged electrons from emitter to collector is equivalent to a conventional current flow in the opposite direction.

The electrical characteristics of the n–p–n transistor are shown in Fig. 5.3. From a simple description of transistor action it would be expected that the collector current would be independent of the voltage V_{CE} when the collector is reverse biased by more than about 300 mV. This is more or less true, although in practice the current does rise slightly with voltage as shown in (a). The transfer characteristic of the BJT (b) shows how the collector current is controlled by the base–emitter voltage V_{BE}.

Both sets of characteristics for the bipolar transistor are reminiscent of those for the MOST (Fig. 4.5) but there are three important differences:

1. The far lower voltage at which the output current I_C of the BJT becomes

Do not confuse this β with the β of the MOST. It is very unfortunate that the same symbol is used for both devices in this series of books and elsewhere. However, the contexts of bipolar and MOS circuits are unlikely to be mixed except when talking about BiCMOS circuits.

The term 'current gain' is misleading. It is commonly used but the reasons are largely historical. The base current is really a loss component. For the MOST, which has zero gate current, the current gain would be infinite!

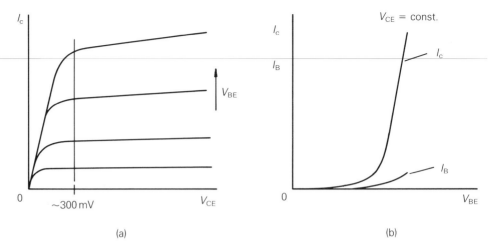

(a)   (b)

Fig. 5.3 The current–voltage characteristics of the n–p–n transistor: (a) the collector (or 'output') characteristics, and (b) the 'transfer' characteristic showing the control of the collector current by the base–emitter voltage.

almost constant ($V_{CE} = 300\,\text{mV}$ compared with ($V_{GS} - V_{Tn}$) for the n-channel MOST);
2. The far more rapid rise of output current with input voltage for the BJT; and
3. The presence of base current in the BJT.

We can use the simple idea of transistor action to derive equivalent circuit models for the BJT which are useful for analysing its operation in a circuit. The d.c. currents and voltages of the BJT are given by two very simple equations describing transistor action. The first gives the collector current of a normally biased device which is approximately equal to the forward current in the base–emitter diode.

$$I_C \simeq I_{ES} \exp(qV_{BE}/kT), \tag{5.3}$$

where I_{ES} is the saturation current of the emitter junction.

The second gives the base current, which is simply

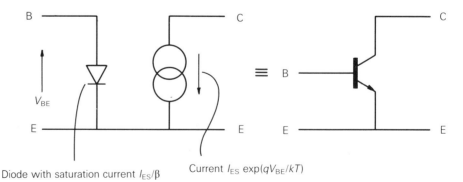

Diode with saturation current I_{ES}/β Current $I_{ES} \exp(qV_{BE}/kT)$

Fig. 5.4 A simple large-signal circuit model for an n–p–n transistor biased in the normal way with the base positive relative to the emitter and the collector even more positive.

$$I_B \simeq I_C/\beta = I_{ES}/\beta \exp(qV_{BE}/kT). \tag{5.4}$$

That is, the base–emitter junction of the BJT is like a normal p–n junction but with the saturation current reduced by the factor β due to transistor action.

These two equations suggest the very simple large-signal circuit model for the BJT with normal bias voltages which is shown in Fig. 5.4. The currents and voltages in this circuit are the same as in (5.3) and (5.4) and it is simpler to use the model rather than thinking about the BJT in a transistor circuit. Although the model is only very approximate, it is adequate for considering the first-order operation of digital circuits.

This is a reduced version of the Ebers–Moll large-signal model of the bipolar transistor and it is quite adequate for simple explanations of the operation of digital circuits.

The Bipolar Transistor as a Switch

The basic circuit for considering the BJT as a switch is shown in Fig. 5.5(a). To see how it operates we can replace the transistor by the approximate large-signal model as shown at (b). We need to find the transfer characteristic of this circuit which is a graph of output versus input voltage.

Ritchie (1987) describes bipolar transistor switching circuits in greater detail.

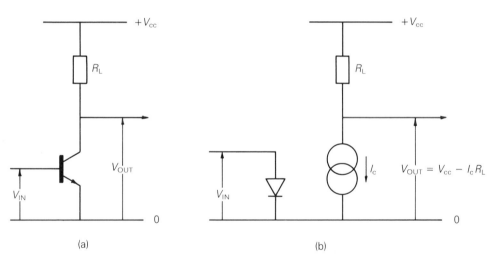

(a)　　　　　　　　　　　(b)

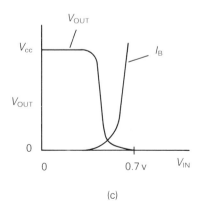

(c)

Fig. 5.5   The bipolar transistor as a switch: (a) the basic circuit, (b) the circuit with the BJT replaced by its first-order model, and (c) the transfer characteristic.

With the input voltage less than about 300 mV the collector current is negligible and the transistor is effectively turned off so that there is no voltage drop across R_L and the output voltage is then equal to V_{CC}. As the input voltage rises to a value of around V_d the collector current rises rapidly and the output voltage falls rapidly as shown in Fig. 5.5(c). We can then say that the transistor has been turned on even though there is no sudden threshold for conduction. Over the same voltage range the base current (equal to I_C/β) also rises rapidly and it has to be provided by the input source. To prevent the base current becoming excessive, the maximum d.c. input voltage has to be limited to about 0.7 V.

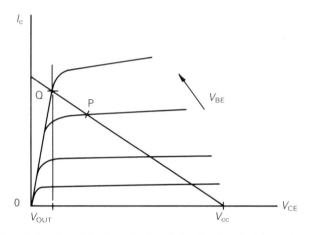

Fig. 5.6 Graphical analysis of the BJT switching circuit.

The use of the word 'saturation' is extremely unfortunate here. For the MOS transistor, 'saturation' refers to conditions giving an almost constant drain current, that is the horizontal part of the characteristic. For the bipolar transistor it refers to the condition when the collector current is saturated with respect to the base current which is the near-vertical part of its characteristic. The nomenclature on the two devices developed independently and it is now so well established that we have to accept it.

The more conventional graphical way of analysing the transistor switching circuit is to draw the load line of slope $1/R_L$ on the transistor I_C–V_{CE} characteristics as shown in Fig. 5.6. This gives the same transfer characteristic but it is more accurate than using the simple circuit model because it includes the effect of **saturating** the transistor by increasing the input voltage from a point such as P to Q where V_{CE} becomes fixed at a value V_{sat} for any further increase of V_{IN}. The value of V_{sat} is typically about 200 mV.

Saturated and Non-saturated Switching

Bipolar digital ICs can be divided into two classes depending on whether or not the transistors are saturated when turned on fully. The most important circuit form using saturated switching is TTL and for non-saturated switching it is ECL, and both of these will be described below. Their properties differ because of the behaviour of the BJT when it is saturated.

When a transistor is driven into saturation the base–emitter voltage V_{BE} becomes greater than the collector–emitter voltage V_{CE} at a point such as Q (Fig. 5.6). This makes the base the most positive terminal on the n–p–n structure so that both p–n junctions become forward biased. Normal transistor action ceases in the saturated transistor because both the emitter and collector junctions inject electrons into the base. The collector current becomes limited by the circuit to the value at the point Q rather than by the base current or voltage.

The advantage of driving a transistor into saturation is that the collector current becomes independent of the BJT properties and it is not affected by manufacturing tolerances in the transistor parameters, in particular β, which cannot be controlled precisely for every transistor on an IC. Saturated switching therefore gives well defined voltage levels in logic circuits but it has a disadvantage in the speed at which a saturated transistor can be turned off. As both junctions of a saturated transistor inject electrons into the base, the electron concentration is far greater than in normal operation. The removal of these electrons, 'stored' in the base, takes a certain period of time, the **storage time**, when the BJT is turned off, making saturated circuits inherently slower. Non-saturating circuits are therefore used for bipolar digital circuits to work at the highest possible speeds and the circuit itself must then ensure that it is not unduly affected by manufacturing tolerances, particularly in the β values.

Transistor-transistor Logic (TTL)

TTL is the most familiar form of logic for standard-product digital ICs. It is also used for ASICs, particularly for semi-custom products using pre-defined library cells. Full-custom TTL design requires considerable skill at the circuit level and is unlikely to be done outside the semiconductor industry so that we need only consider the basic operation of a typical circuit as a background to the use of library cells.

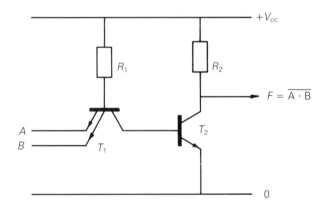

Fig. 5.7 A basic two-input TTL NAND gate circuit.

The circuit of a two-input TTL NAND gate forming an internal part of an IC is shown in Fig. 5.7. It contains two transistors, T_1 and T_2, the first of which is a truly integrated component with two n^+ emitter regions in a common p-type base. Transistor T_2 is a straight switch that is effectively an open circuit when it is OFF and with only the voltage V_{sat} across it when ON, which is equivalent to logic 0.

The inputs A and B are nearly always connected to the outputs of similar gates so that they are linked to V_{CC} through the resistances R_2 of previous stages and to ground via the T_2 switches which are open for logic 1 and closed for logic 0.

To see how the circuit works we start by considering the effect of changing the input A, while B remains at logic 1 (Fig. 5.8). When A is also at logic 1, effectively disconnecting it from ground, current flows through R_1, the base–collector junction of T_1 and the base–emitter junction of T_2 as shown at (a). The collector of T_1 acts like an emitter in this condition with a voltage drop of V_{BE}, and the

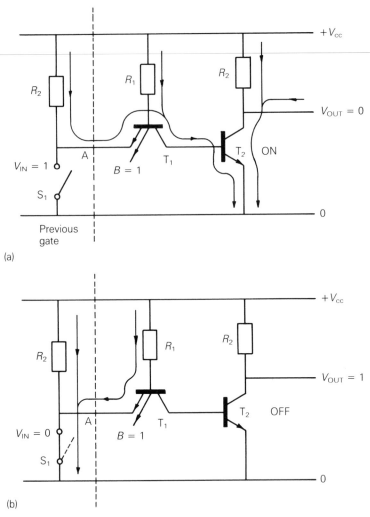

(a)

(b)

Fig. 5.8 The basic TTL circuit redrawn and including parts of the previous circuit and the load: (a) shows the current flow for $A = B = 1$, $F = 0$, and (b) is for $A = 0$, $B = 1$, $F = 1$.

'emitter' at A acts as the collector since it is reverse biased. The current through A is limited by R_2 of the previous stage because the circuit is designed to saturate T_1. The input voltage at A is therefore made up of V_{sat} across T_1 and the base–emitter voltage of T_2. This is approximately $(0.2 + 0.6)$ volts, which is the logic 1 voltage in the circuit. As T_2 is turned on, current also flows through R_2 and the output is at logic 0.

If the input A is now changed to logic 0, the current paths change as shown in Fig. 5.8(b). Logic 0 is represented by closing the switch S_1 and in this condition there is insufficient voltage at A to turn on the current path through T_1 and the base–emitter junction of T_2. T_2 therefore remains off and the base current of T_1 is diverted through A and S_1 to ground together with the current through R_1. As T_2 is off the output node F is pulled up to a logic 1 voltage by the following stage. The circuit therefore gives the inverse of the input state when B is fixed at logic 1.

It requires a voltage $V_{IN} = V_{sat} + V_d$ to turn on this path. The logic 0 voltage is only V_{sat}.

We finally consider the conditions with B at logic 0 which again diverts the base current of T_1 to ground, but this time through the input B. T_2 is therefore off and the output is logic 1. Summarizing, it can be seen that if either A or B or both are at logic 0 the output is logic 1. The output is at logic 0 only when both A and B are at logic 1 and this defines the NAND function, $F = \overline{A \cdot B}$. The principle of this circuit can be extended to multiple inputs simply by adding more n^+ emitter regions to T_1.

The multi-emitter transistor in TTL has an almost constant base current and the difference between the conditions (a) and (b) in Fig. 5.8 is that it is diverted from the 'collector' (really the emitter in terms of a conventional transistor) to an emitter if one or more of the inputs go to logic 0. Hence, although T_1 is saturated, it is never completely turned off so that the switching delay is determined largely by T_2 which is switched off when the output changes from logic 0 to 1.

A NOR circuit is made in TTL by paralleling the T_2 transistors of two inverters. Combinations of ANDs and NORs can also be made in single circuits, as for example in Fig. 5.9. All logic functions and bistables are built up from NAND and NOR gates in TTL cell libraries.

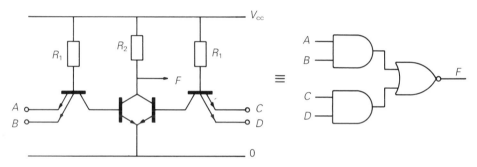

Fig. 5.9 A TTL-type circuit for the AND–OR–NOT function.

There are many designs of TTL circuit that give rise to the different families of standard digital ICs. The circuits themselves differ in the values of R_1 and R_2 and in the detailed design of the transistors, and in some of them R_2 is dispensed with entirely and/or extra diodes may be added to the circuit. Details can be found in more advanced textbooks.

The circuit of Fig. 5.7 is for a gate as an internal part of an IC. However, more powerful circuits are needed for driving signals off the chip and a push–pull output stage is normally employed in TTL pad driver circuits.

See, for example, Elmasry (1983)

Work out the logic voltages V_{HI} and V_{LO} for a TTL circuit in which R_2 is omitted. Take $V_{sat} = 0.2\,\text{V}$ and V_{BE} for an ON transistor to be equal to $V_d = 0.6\,\text{V}$.

Exercise 5.3

Calculate the currents taken by a TTL circuit in which $R_1 = 10\,\text{k}\Omega$, $R_2 = 4\,\text{k}\Omega$, $V_{CC} = 5\,\text{V}$, when the inputs are at both logic 1 and logic 0 using the same parameters as in the previous exercise. Hence evaluate the average power dissipation of the gate.

Exercise 5.4

Schottky TTL

The speed of TTL can be improved significantly by the very simple addition of Schottky diodes. The Schottky diode is a rectifier which is formed at the interface between n-type silicon and a metal, typically an alloy of aluminium. It has approximately the same $I–V$ characteristic as the p–n junction but with the important difference that the value of I_s is several orders of magnitude greater than for a p–n junction of the same area made on the same n-type silicon. The forward voltage drop of a Schottky diode for a 'reasonable' current (for example, $100\,\mu A$) is therefore considerably reduced, perhaps to 0.35 V compared with 0.6 V for the p–n junction.

The second important difference between the Schottky diode and the p–n junction is that the forward current is carried by the flow of electrons from the silicon to the metal, where they almost immediately 'mix' with the very large number of electrons always present in metals. If the voltage on a Schottky diode is suddenly reversed, there is hardly any delay in turning off the current because there is nothing equivalent to the extraction of the injected electrons that slows down current changes in p–n junctions.

The Schottky diode is used in an extremely ingenious way to improve the switching speed of transistors in TTL. The addition of a Schottky diode between the base and collector of an n–p–n bipolar transistor is known as a **Schottky clamp**. We have seen that the base–collector junction of the transistor becomes forward biased when it is strongly saturated, and when this occurs the forward voltage can rise to about 0.6 V. However, the Schottky diode connected in parallel with the junction conducts heavily when the forward voltage gets to about 0.35 V and this prevents the base–collector voltage rising further so that the transistor cannot become deeply saturated. A clamped transistor therefore has far fewer electrons stored in the base when it is conducting heavily so that it can be switched off far more rapidly.

A Schottky clamped transistor. Note the circuit symbols used.

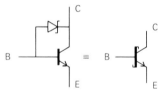

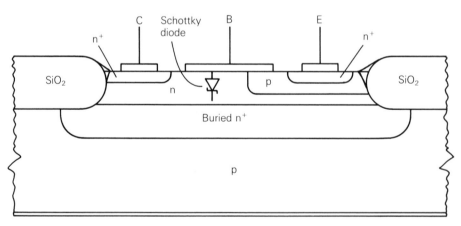

Fig. 5.10 Section of an isolated n–p–n transistor with a Schottky clamp.

The Schottky diode is incorporated into the device structure by the very simple modification of extending the base metallization of an n–p–n transistor over the collector (Fig. 5.10). The metal makes an ohmic contact to the p-type base but a rectifying contact to the n-type collector region. The collector contact itself is ohmic because of the heavy n^+ doping put in below the metal.

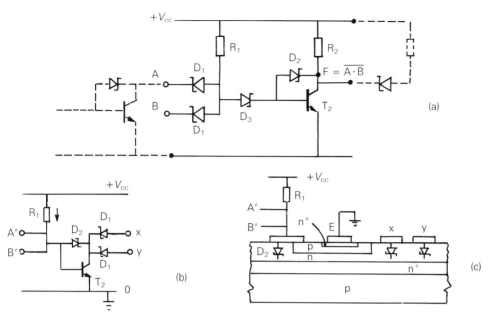

Fig. 5.11 (a) An STTL circuit in which T_1 is replaced by Schottky diodes. (b) The circuit of STL and (c) its structure.

In Schottky TTL (STTL), T_2 is clamped in this way and T_1 may also be. However, Schottky diodes may be used to replace T_1 entirely in some designs because of their higher switching speed (Fig. 5.11). The circuit is then reminiscent of diode transistor logic (DTL) which was the earliest type of digital IC, using p–n junction diodes, and which has long been obsolete. STTL is, of course far faster than any DTL circuits.

Calculate the logic voltages for the STTL circuit shown in Fig. 5.11. Use the same transistor parameters as in Exercise 5.3 and take the forward voltage drop of the Schottky diode as 0.35 V.

Exercise 5.5

Other Related Circuit Technologies

There are many other ways in which digital functions can be implemented with bipolar transistor switches. The objective in developing these has been to make compact circuits to operate at high speeds with reduced power consumption for VLSI applications.

One of these developments, Schottky transistor logic (STL) is easily derived from STTL which was shown in Fig. 5.11(a) with the input and output connected to similar gates. The circuit is redrawn in Fig. 5.11(b) with three changes.

1. The boundaries between one stage and the next have been shifted to the right of the diodes D_1,
2. Resistance R_2 and diode D_3 have been omitted, and
3. Schottky diodes D_1 and D_2 are now made with different metals to given different forward voltage drops V_{S1} and V_{S2}.

Typically, titanium used for D_1 and aluminium for D_2 would give $V_{S1} = 0.3\,V$ and $V_{S2} = 0.5\,V$. If V_{BE} (= V_{H1}) is 0.6 V, V_{L0} is ($V_{BE} - V_{S2} + V_{S1}$) = 0.4 V. The logic swing is then 0.2 V.

See, for example, Hodges and Jackson (1983).

This is now the circuit of STL. It operates in a similar way to the earlier TTL circuit. If one of the inputs is low, all the current I flows through its input diode and T_2 is off. The transistor sinking the current is saturated so that the voltage V_{LO} at the input A' is ($V_{sat} + V_{S1}$). If both inputs are high, I flows through the base–emitter junction of T_2 and the voltage V_{HI} at A' is V_{BE}. T_2 is turned on and its collector voltage is clamped at V_{sat} equal to ($V_{BE} - V_{S2}$) by Schottky diode D_2. The logic swing ($V_{HI} - V_{LO}$) is then just the difference between the Schottky diode voltages V_{S2} and V_{S1}. The circuit now operates perfectly well without the R_2 and D_3 of STTL so that they can be omitted.

An important advantage of STL is that the entire circuit can be integrated into a single device structure (Fig. 5.11(c)). The Schottky diodes are made by appropriate metallizations to the n-type collector material of T_2 and there is no longer any need for a separate collector contact. Three or four output diodes D_1 can be made to fan out to other inputs. A single resistor R_1, integrated on the chip, can supply the current to many gates.

A related circuit form, **integrated injection logic**, I²L, uses normal collector junctions instead of Schottky diodes for the outputs and a constant current source made with a p–n–p transistor to replace R_1. The p–n–p and n–p–n transistors can be merged into a single structure that is nearly as small as an STL gate.

STL and I²L consume less power than other bipolar circuits, and STL can be faster than TTL. At the present time they have been rather overshadowed by advances in CMOS and they are not widely used. They are mentioned here, however, to make the reader aware that there are many possibilities for alternative technologies in the future.

Emitter-coupled Logic (ECL)

To obtain the highest possible switching speed in bipolar digital circuits it is essential to prevent the transistors becoming saturated. It is then necessary to devise a circuit form in which the current is more or less independent of the gain of the transistors because β cannot have a value that is precisely reproducible either between the transistors on a chip or, even less, across the whole area of a wafer.

Circuits based on the differential amplifier or **long-tailed pair** configuration (Fig. 5.12(a)) are of this type. In this circuit the total current in the two branches I_0 is held almost constant by a constant-current element CCE which may itself be a transistor circuit or a high-value resistance. If V_A and V_B are equal and the circuit is perfectly symmetrical, the current I_0 divides equally between the two branches, so that V_P and V_Q are also equal. If V_A is made greater than V_B, the current in the left-hand branch rises while that in the right-hand one falls so that the voltage at P goes more positive and Q goes negative. Within limits the circuit responds only to differences between V_A and V_B.

By replacing the transistors by their circuit models (Fig. 5.4) it can be shown that the output voltages of a differential amplifier change with the differential input as shown in Fig. 5.12(b). For ($V_A - V_B$) greater than about $\pm 4kT/q$ (approximately $\pm 100\,mV$), all the current flows in one branch and the outputs are at either V_{CC} or ($V_{CC} - I_0 R_L$). For digital applications the outputs can therefore be set in to either one state or the other corresponding to logic 1 and logic 0 by applying a difference of more than about $\pm 100\,mV$ to the inputs.

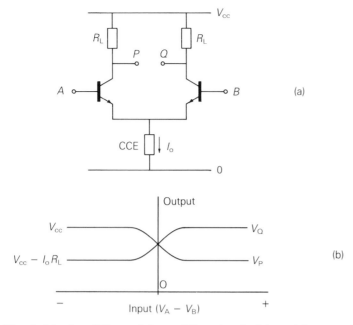

Fig. 5.12   A bipolar differential amplifier circuit (a) and its transfer characteristic (b).

This circuit is the basis of emitter-coupled logic (ECL) which is used for very high-speed digital ICs. The design of all bipolar circuits requires very detailed analysis which makes full-custom design a specialized activity. However, cell libraries remove these problems and they are widely available for ECL semi-custom ASIC design, so it is worthwhile learning something about how the circuits operate.

The principle of a two-input ECL NOR gate is shown in Fig. 5.13. As shown, it uses a resistor R_S to give an approximately constant current I_0, although a transistor circuit is often used in practice. The right-hand branch contains a

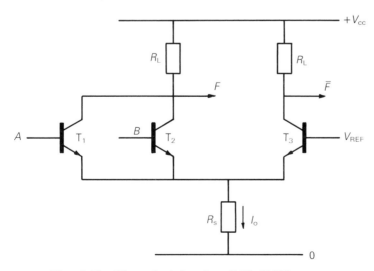

Fig. 5.13   The principle of an ECL NOR gate.

83

single BJT with the base at a fixed d.c. voltage V_{REF}, and the left-hand branch has two transistors in parallel with the inputs A and B connected to their bases. If the voltages at A and B are both low (logic 0) nearly all the current I_0 flows through T_3 so that there is hardly any voltage across R_L and the output F is high (logic 1). If either A or B or both go high the current is diverted to the left-hand branch and F goes low (logic 0). Hence F gives a NOR function of A and B. ECL circuits usually give an inverse output also: here $\bar{F}$ is the OR function of A and B.

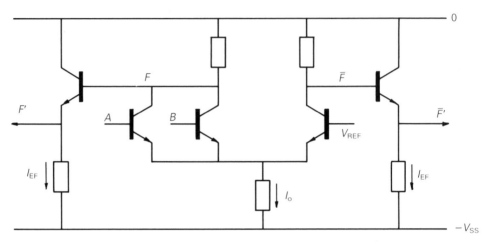

Fig. 5.14   A complete ECL NOR gate circuit.

Because ECL is very fast, great attention has to be given to driving the load external to a gate and this is usually done by adding low-impedance emitter follower circuits to both outputs as shown in Fig. 5.14. The power supply is usually negative with ECL because the logic voltages are close to the positive rail. Careful electrical design is needed to give logic 1 and 0 voltages that can be connected directly from outputs to inputs while ensuring that the transistors are not saturated. Values of about 0.2 V on either side of V_{REF} can be used on a chip.

See, as before, Hodges and Jackson (1983).

Although ECL circuits are extremely fast they consume a lot of power because of the need for the steady currents I_0 and I_{EF} (Fig. 5.14) which can be several hundred microamps per gate. The current can be used more effectively by 'stacking' the logic transistors in two or more levels as shown, for example, in Fig. 5.15. This requires different reference voltages V_{REF_1} and V_{REF_2} for each level and extra circuits to d.c. shift the logic signals for connection to level 2 inputs. Complex functions and bistables can be built up in stacked ECL.

Current Mode Logic (CML)

There are many variants of the non-saturated switching principle used in ECL. Current mode logic (CML), is similar to the ECL circuit described above but with the emitter followers omitted so as to reduce the power consumption and circuit size, but at the expense of circuit speed. Another variant which is also called CML has been widely used in the semi-custom ICs manufactured by Ferranti Semiconductors (now part of Plessey Semiconductors).

Ferranti CML uses transistors made with a modification of junction isolation known as **collector-diffused-isolation** (CDI). This combines the isolation ring

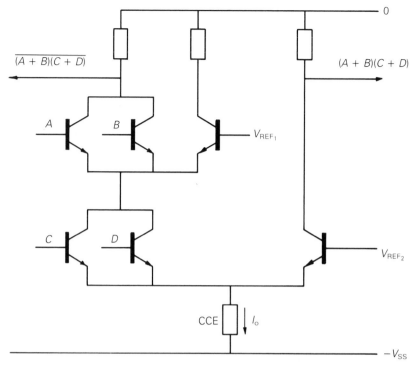

Fig. 5.15   A stacked ECL logic circuit.

with the collector contact as shown in Fig. 5.16 which should be compared with the standard junction isolation using a non-active isolation ring (Fig. 2.10). The CDI transistor uses a p-type epitaxial layer as the base region of the n–p–n structure so that the isolation ring can be n^+ and, by diffusing it right down to the buried layer, it can form the collector also. The resulting transistors are moderately small and particularly simple to make but they only work at the low voltages used in the Ferranti version of CML.

The basic circuit used in this form of CML is the two-input NOR gate shown in Fig. 5.18. The main difference compared with ECL is that no voltage reference is required because the circuit is more like a current-limited switch than a differential amplifier.

This CML circuit form can use a very low on-chip supply voltage of 0.92 V. If

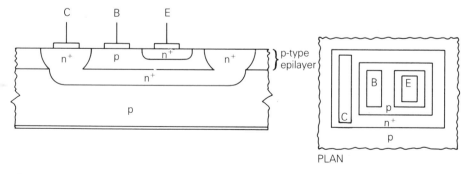

Fig. 5.16   A collector-diffused-isolation n–p–n transistor. (Plan view to reduced scale.)

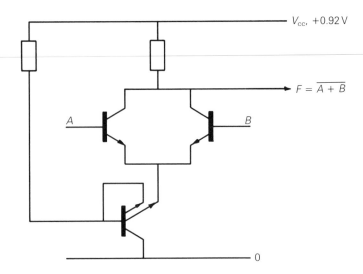

Fig. 5.17 The basic CML two-input NOR gate circuit.

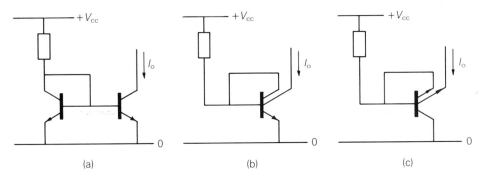

Fig. 5.18 The basic current mirror circuit showing at (b) how it is integrated into a single-element current source, and redrawn in the CML form at (c).

either or both of A and B is at logic 1 (0.92 V) at least one of the transistors conducts and the output is low (0.56 V). The output is only high when both transistors are switched off by applying logic 0 to both A and B, and this defines the NOR function. When the circuit is conducting the current is limited by the multi-emitter transistor which is a current mirror (Fig. 5.18) but with the collector and emitter symbols reversed as in (c) to represent the particular way in which the device is fabricated.

The basic CML circuits of this type have been developed in the same ways as ECL, with variants containing emitter followers and stacked logic transistors (called **differential CML**).

Comparison of Circuit Technologies

In order to choose the most appropriate technology for an IC, the designer must have a good understanding of the relative merits of the different circuit forms. The criteria for comparison were described early in Chapter 4 as speed, power,

The operation of the current mirror is described in Ritchie (1987).

and packing density for gate circuits, and they can be combined into power-delay and packing density-frequency figures of merit.

It is extremely difficult to make a quantitative comparison across the entire field because of

1. the wide range of circuit designs and fabrication technologies used within each circuit family by different comparies,
2. the rapid rate at which circuit technologies are changing, and
3. the difficulty in collecting information in a uniform way.

The third problem arises because the performance of circuits depends on their precise design and application. Ideally one should compare the delay, power dissipation and area figures for minimum-sized gate circuits with a stated fan-out. This information is often provided for cell libraries but for other circuits only the output characteristics of complete ICs may be available.

In spite of these problems, the differences between the circuit technologies are sufficiently great to allow some generalizations to be made. First, in comparing bipolar and MOS circuits, the differences are due to the basic properties of the two types of transistor which were listed on pages 73–74. Electrically, the bipolar transistor is superior to the MOST for two reasons.

1. The current is almost constant down to a very low collector voltage (Fig. 5.19(a)), and
2. The control action is extremely sensitive, $I_C \propto \exp V_{BE}/kT$, compared with $I_d \propto (V_{GS} - V_T)^2$ for the MOST (Fig. 5.19).

As a consequence, the bipolar transistor has a better current-driving capability for charging capacitive loads at high speed although it also has the additional delay due to charge storage. The BJT also the disadvantage of requiring input (base) current which results in more complicated circuit forms, usually with resistors as well as transistors, and relatively high power dissipation. We have already seen that the power dissipation of individual gate circuits ultimately limits the maximum packing density that can be achieved on an IC because of the practical limit to the removal of heat from the chip. For example, if a bipolar gate dissipates $200\,\mu W$, it will only be possible to run about 2500 of them

The current cannot be reduced further in the design of digital bipolar circuits because the gain of BJTs falls off at low currents. As it becomes possible to make smaller transistors, however, the current can be reduced, keeping the current density constant.

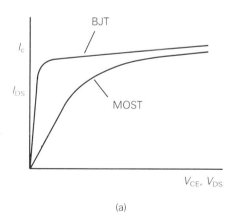

(a)

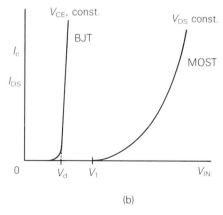

(b)

Fig. 5.19 The important differences between the characteristics of bipolar and MOS transistors: (a) the output characteristic, and (b) the transfer characteristic.

simultaneously on a single piece of silicon without it becoming unreliable due to overheating, unless special cooling methods are employed.

Bipolar circuits are generally regarded as being more 'rugged', referring to their reliability in harsh operating environments both physically and electrically. The MOS transistor depends on the insulation of the very thin gate oxide which can be destroyed by excessive voltage spikes in spite of protection built into input circuits of MOS ICs which may reduce their reliability in particularly demanding applications.

The properties of the main circuit technologies for ICs can be summarized as follows.

1. TTL is moderately fast with gate delays of less than 1 ns in some products. Very reliable but limited to low LSI levels of integration by power dissipation.
2. STL is moderately fast and it has the high packing density and low power needed for VLSI. I^2L has similar packing density and it can take less power than STL but it is considerably slower.
3. ECL is very fast with internal gate delays of down to 100 ps but with relatively high power per gate which limits the maximum packing density to a few thousand gates per chip. ECL ICs are the fastest yet produced and they are used for very high-speed logic such as in the central processing units of main-frame computers.
4. Both forms of CML are slower than ECL but, with lower power, many thousand gate equivalents can be integrated on a single chip.
5. n-MOS is moderately fast with internal gate delays of less than 1 ns. The cell size is small. The static power dissipation is a disadvantage compared with CMOS and this will ultimately limit its use for VLSI.

1 ps = 1 picosecond = 10^{-12} s. Light travels about 30 mm in a picosecond.

n-MOS

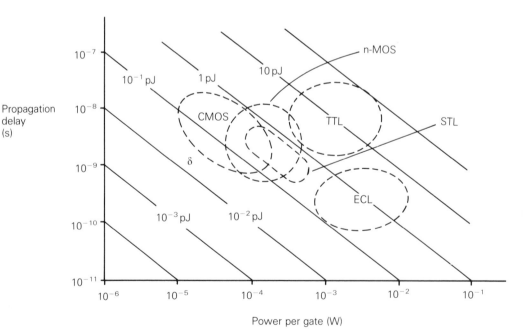

Fig. 5.20 The approximate positions of the main integrated circuit technologies on a delay-power plot. The MOS circuits are assumed to be operating at their maximum frequency.

6. CMOS is moderately fast and it has a small cell size when multi-layer metal-lization is used. Otherwise not quite as compact as n-MOS. The great advantage of its zero static power dissipation makes it the only technology for advanced VLSI products although, as circuits are made smaller and faster, the packing density will ultimately be limited by the dynamic power dissipation which is proportional to frequency.

The properties of different technologies are sometimes compared by showing their approximate positions in the **delay – power plane** as in Fig. 5.20. The diagonals on this double-logarithmic plot are lines of constant power-delay product and they can be associated with switching energy in picojoules. However, such diagrams have to be treated with great caution because the regions shown for each technology may not be for identical conditions and there may be differences in the way that static and dynamic power values are combined.

Combined Bipolar/CMOS Technology (BiCMOS)

The main advantage of CMOS is its high packing density and low power dissipation. However, its low current-drive capability increases the delays in charging and discharging the larger capacitive loads on an IC such as the clock lines and tracks for global control signals, and this can slow up the operation of the entire chip. The idea behind the combined bipolar/CMOS technology known as BiCMOS is to use the larger driving capability of bipolar transistors to reduce the delays in driving such relatively large loads, and possibly small ones also in some designs.

A basic BiCMOS circuit is shown in Fig. 5.21. It is similar to a pure CMOS inverter but with the addition of two resistors R_1 and R_2, and two n–p–n transistors. When the circuit is in equilibrium with either a steady 1 or 0 on the input, both bipolar transistors are off because there is no current flow through either resistor and hence no voltage between the bases and emitters. However, when the circuit changes state from 0 to 1 at the output, for example, the current

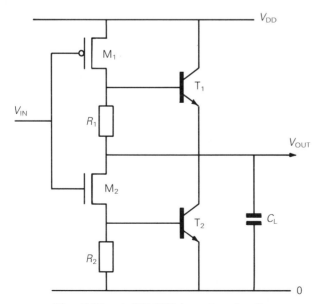

Fig. 5.21   A BiMOS inverter circuit.

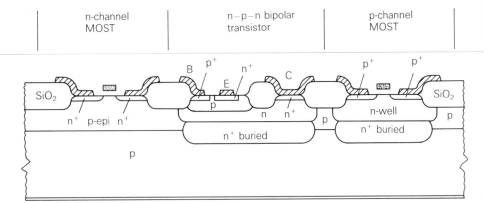

Fig. 5.22 Addition of an n–p–n transistor to a CMOS process for BiCMOS circuits. The buried n$^+$-layer, required for the n–p–n, is also put under the p-MOST to improve the resistance to latch-up.

charging C_L flows through R_1 developing a voltage to turn on T_1 which therefore supplies additional current to C_L, charging it more rapidly than in a normal CMOS circuit. When C_L is fully charged, T_1 is turned off again as V_{BE} falls. The sequence is similar when the output discharges, with T_2 being switched on for a short period only. As in CMOS there is no static power dissipation.

The fabrication of high-speed n–p–n transistors in the same piece of silicon as the CMOS circuit requires two additional steps as seen in Fig. 5.22. The n–p–n transistors are made in the CMOS n-wells but they require the diffusion of an additional p-type layer to form the base. The emitter and the base contacts of the BJTs are formed at the same time as the n$^+$- and p$^+$-MOST sources and drains, but the buried n$^+$-layer, needed to get good bipolar characteristics, is the second addition to a normal CMOS process.

BiCMOS technology can substantially improve the speed of a VLSI circuit but at the expense of the extra fabrication stages which may limit its large-scale use to special applications such as static memory chips. It also has considerable potential for use in high-frequency mixed analogue–digital ICs.

Summary

Although CMOS is widely used for ASICs at present it is important for designers to understand the principles of bipolar transistors and circuit technologies in order to select the best one for any application. This chapter has described the main circuit forms used in bipolar digital ICs. They use either saturated or unsaturated switching. The speed penalty with saturated switching is reduced by adding Schottky diodes to circuits such as TTL. Non-saturated switching is employed in ECL and various forms of CML which are generally faster.

The bipolar transistor inherently has a higher gain than any form of MOSFET and its collector current remains almost constant to lower voltages as shown in Fig. 5.19. These properties are used to make the fastest digital circuits (ECL) and other forms that trade off some of the speed for power or packing density. However, the relatively high power consumption of all bipolar circuits prevents them being used in the most complex ICs.

ASIC and Programmable IC Technologies   6

By studying this chapter you should be able   **Objectives**
- [ ] To understand the importance of ASICs for the production of electronic systems and their benefits.
- [ ] To appreciate the advantages of the standard cell and gate array methods for the implementation of ASIC designs and their relative merits.
- [ ] To understand the reasons for the cost and performance differences between standard cell and gate array chips in order to choose the best approach to use for a particular ASIC.
- [ ] To understand the basic technologies used in electrically programmable ICs in order to compare them with mask-programmed ICs.

In this chapter we will describe some of the special technologies used for application-specific integrated circuits (ASICs). An ASIC is a chip designed for use in one particular electronic system and the total number of chips required may be relatively small, perhaps only a few thousand compared with the millions for a standard-product IC. To make ASICs at all economically it is therefore necessary to reduce the costs of making small numbers. This can be done by

1. Reducing the design cost, and
2. Reducing the fixed manufacturing costs.

The design time, and hence the cost, is reduced by the combination of semi-custom methods and CAD that will be described in Chapter 8. Semi-custom design uses a library of pre-defined cells that, with the appropriate CAD tools, can be assembled rapidly to create the design of a complex IC. Although this eliminates the need for detailed circuit and layout design, we will find that it also constrains the chip architecture to a certain extent.

The manufacturing cost is reduced either by cutting down the number of masks needed to fabricate a particular design as a **masked IC**, or by using a **programmable IC** which is a standard-product chip that is electrically configured by the user for a particular application. This chapter is concerned with both these technologies but, before describing them in more detail, we need to see why ASICs are so important for modern electronics.

Use of ASICs in Electronic Systems

ASICs have many advantages over standard-product ICs for use in electronic systems. The most important ones are:

1. Reduction in the total number of chips in the system because a single ASIC can easily replace a whole board of standard ICs;

One of the problems of using standard ICs is that they are seldom exactly right for the particular application, for example in the control signals or register size. Extra **glue logic** therefore has to be added to make up for the deficiencies. A single ASIC, however, can be designed to do the complete job in any way required for the particular product, and it may easily add further functions at little extra cost.

2. Higher performance because the design can be optimized for the single application;
3. Ability to add extra functions to an ASIC at little extra cost;
4. Greater security of the design which is necessary since a board of standard ICs is very easily copied by an unscrupulous competitor.

The disadvantages of ASICs are often said to be high design cost, long lead-time for fabrication, and the high cost of making design changes subsequently. Semiconductor companies are continually making improved products and CAD systems for reducing these difficulties in a very competitive market. In particular, programmable and reconfigurable ICs, which enable the designer to produce prototypes himself, are used increasingly for the smaller ASIC functions.

The advantages of increasing the level of integration of ICs by packing more circuitry on to single chips apply just as much to ASICs as to standard-product ICs. They include lower total silicon cost, higher operating speed and reduced power consumption. The use of complex ASICs may therefore lead to further advantages in the production of electronic equipment and systems. Some of these are:

1. A reduction in the number of printed circuit boards in the complete system with obvious savings in size and weight.
2. Increased reliability due to the smaller number of exposed connections which are the main cause of failure in modern electronics.
3. Reduced total power consumption and hence a smaller power supply.
4. Increased operating speed due to a smaller number of off-chip connections which reduce the flow of digital data.

The reductions in size, weight and power lead to further savings in the cost of enclosures and cooling. For some applications, such as in avionics, they are absolutely crucial.

These advantages explain why new electronic ystems tend to contain smaller numbers of more complex ASICs each year. By understanding the design and manufacturing methods we can see how the cost of ASICs can also be reduced so that they are becoming some of the most commonly used components in the whole of electronics. We consider manufacturing methods first because they indicate what it is that we have to design. Semi-custom design will be described in detail in Chapter 8.

Standard-cell and Gate-array ASICs

Semi-custom design methods based on a library of predesigned circuit cells are used for ASICs that will be produced as **standard-cell** or **gate-array chips**. These terms refer more to chip architecture than technology although the design methods and the mask and chip fabrication techniques become inseparable when considering ASICs. We will consider these semi-custom chip architectures as used for both MOS and bipolar digital ICs in this chapter and will add a little on the inclusion of analogue functions in Chapter 10.

Standard-cell ICs

The cell libraries used in semi-custom design contain data on the functions of all cells likely to be needed in digital systems including gates with varying numbers of inputs, bistables and larger functions such as adders or registers. The type of information given in the library will be described in more detail in Chapter 8. For chip layout it includes the dimensions of the cells for implementation by a particular fabrication technology. The cells are usually rectangular, with a fixed height but variable width depending on the size of the circuit they contain, and the inputs, outputs and power supply connections are at stated positions on a fixed grid round the edge of the cell. The inputs and outputs are often duplicated on the top and bottom edges to simplify the problems of connecting them.

In most standard-cell ICs the individual cells are arranged in rows separated by **routing channels** for the interconnections as shown in Fig. 6.1. This regular architecture simplifies the task of laying out the connections in accordance with the logic circuit of the IC. The paths of the connections are usually determined automatically using an **autorouting program** which is included in the CAD system. The interconnections consist of horizontal and vertical metal tracks as shown in Fig. 6.2. and the CAD program may adjust the widths of the routing channels depending on the number of interconnections they contain so as to minimize the

The orientation of an IC is, of course, purely arbitrary. Rotating Fig. 6.1 by 90° turns rows into columns and at least one manufacturer uses this notation!

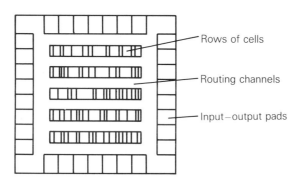

Rows of cells

Routing channels

Input—output pads

The input and output 'pads' are metallized areas for the attachment of the external wires with circuits for interfacing between the chip and the outside world.

Fig. 6.1   Layout plan of a typical standard-cell IC.

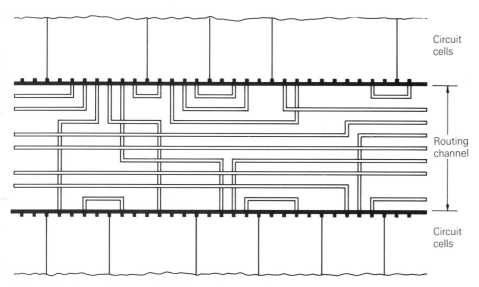

Circuit cells

Routing channel

Circuit cells

Fig. 6.2   A small part of the layout in the routing channel of a standard-cell IC showing typical double-metal interconnections with vias at the corners.

As an alternative to double-metal tracks the vertical tracks may be made of polysilicon. The power supply connections are generally run through the cells (as shown in Fig. 6.3 and 6.4) so as not to impede the routing channels.

An example of a chip plan designed partly from cells combined into blocks and partly from rows of standard cells.

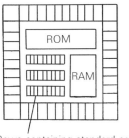

Rows containing standard cells

This is called *e-beam direct write*.

The term 'gate array' is strictly incorrect. The chip contains an array of components from which gates can be made so it should be called a 'transistor array'. The original term used by the originators of gate arrays, Ferranti Semiconductors, was 'uncommitted logic array' (ULA) which is much more accurate.

chip area. Where one track is required to cross over another, connections are made to an **underpass** which is a short track at a lower level, very often made of polysilicon in MOS ICs. There is, of course, no need for underpasses if two layers of metallization are used as in many fabrication processes.

The chip architecture in standard-cell design is determined by the type of CAD system used. With the simpler systems, the cells are arranged in regular rows as in Fig. 6.1, but with more advanced software tools, the system can assemble cells into arrays before determining their positions on the chip plan. The arrays are often large blocks such as ROM, RAM, or PLAs, as well as counters, registers, etc., of varying size. They are usually composed automatically from a description of the design written in a hardware description language as will be described in Chapter 7. The positions of the blocks, fitted on to a grid within the chip plan, and the routing of the interconnections is then determined automatically to complete the layout. This style of design gives a higher electrical performance and a more compact chip and it is used increasingly for designing complex ICs in the minimum possible time.

Use of standard cells reduces the design time dramatically but it has no effect on the cost of manufacturing small quantities because a full set of masks for the layout still has to be produced for fabrication. Although the internal details of the cells are not available to the designer, the layout of every cell is added when the masks are made for a specific design. The cost of a full set of perhaps 12 masks may prohibit the use of the standard-cell approach if the number of chips to be produced is extremely small.

At least one company, European Silicon Structures (ES2) overcomes this by fabricating ICs without the use of conventional masks by exposing the photoresist layer on the wafer with an electron beam. The electron beam writes the required pattern at every stage of lithography under the control of a computer which contains the layout information. Although this is an ideal method for making extremely small numbers of chips, the high cost and low throughput of electron beam equipment makes it uneconomical for large-scale production so that conventional masks are made if a large numbers of chips is required after the prototypes have been found to work successfully.

Gate Arrays

Gate arrays use predefined cells to reduce the design cost as in the standard-cell approach. However, they go one stage further in making low-volume ASICs economical by also reducing the manufacturing cost. They do this by reducing the number of masks needed for the fabrication of a particular design from perhaps twelve for a standard-cell IC to three or even one. The ability to do this rests on the fact that all the cells can be designed using the same identical transistors and, in bipolar circuits, resistors, that are then connected up in various ways to produce the different functions. The basic components are formed in the lower levels of an IC and, in a gate array, they are fabricated in repeating groups that are connected only at the upper levels to implement any number of designs. Wafers containing gate-array chips can therefore be mass produced at low cost up to the metallization stage and they are **committed** to a particular design only in the final stages of fabrication. This can reduce the fixed (mask) cost of fabrication by up to 75%.

The most commonly used chip architecture for gate arrays is similar to that for standard cells (Fig. 6.1) with the component groups fabricated in rows leaving routing channels between them. The only difference from the standard-cell architecture is that the routing channels have to be of constant width since the lower levels of the IC are already at fixed positions on the chip. The channels may also contain a large number of regularly spaced underpasses whereas in a standard-cell IC they are only fabricated where they are needed.

The component rows contain regularly repeating groups of transistors and, in bipolar systems, resistors. The basic element of a CMOS gate array typically contains two n-channel and two p-channel transistors connected as shown in Fig. 6.3. Extra connections are added to make up the various cell functions as shown, for example, in Fig. 6.4 for a three-input NAND gate made using two basic elements. A more complex functions would use several elements, built up into a larger cell with the size given in the library information. The actual internal connections of the cells are also sometimes given in gate-array data books even though they cannot be changed by the designer.

Gate-array wafers are fabricated in bulk up to and including the deposition of the metal over the entire wafer. All the cell contact windows will previously have been etched out so that the metal makes contact with all the devices. To commit the wafer to chips of a particular design, the metal is etched away to leave the required connections both inside and between the cells. With a single layer of metallization this uses only one stage of lithography and only one mask for the complete design.

Double-layer metallisation is used for most of the more advanced gate arrays. In committing them, the first layer of metal is etched to the required pattern and it is then covered by a continuous insulating film. The second mask for the particular design is used to define the contact holes in this film to form vias after which the second layer of metallization is deposited. The interconnections are then completed by etching this film as defined by the third mask for the design.

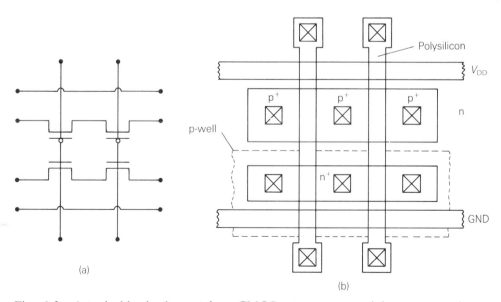

(a)

(b)

Fig. 6.3   A typical basic element for a CMOS gate array containing two n- and two p-channel transistors: (a) circuit diagram, and (b) layout. The V_{DD} and ground rails for this design are in Metal 2.

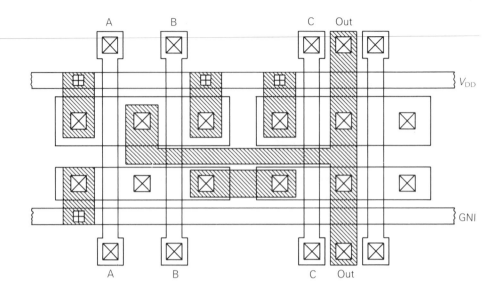

Fig. 6.4 The Metal 1 connections (shaded) that are added to two basic elements like Fig. 6.3 to form a three-input NAND gate. The contacts with vertical crosses are vias between the two levels of metal as shown in Fig. 2.14.

Although three masks are used, two layers of metallization can produce a far more compact layout, and hence a smaller chip, than single metal.

Compared with the circuits used in the standard-cell method, gate-array cells have a lower performance because all the transistors are identical. In Chapter 4 we saw that in the design of CMOS circuits the transistor widths should be adjusted to give the best transient response. However, it is not possible to change transistor sizes in gate arrays so that the circuit is likely to be slower than if its electrical design was optimized. Wider transistors are also needed for buffer circuits to drive large loads, and in gate arrays they are made by paralleling two or more minimum-sized inverters instead of adjusting the transistor dimensions.

One of the main differences between gate-array and standard-cell ICs is that a gate-array implementation of a design has to be fitted into a fixed chip size. Manufacturers produce 'families' of gate arrays containing different numbers of the same basic elements, for example, 600, 1200, 2400, 6000, and 10 000 elements per chip with correspondingly increasing numbers of I/O connections. A design has to be fitted in to the smallest chip of the chosen family that will take it and this may leave many of the circuit elements unused. The **chip utilization** is the percentage of the elements used by a particular design, figures between 60% and 85% being fairly common. In a standard-cell IC, on the other hand, the utilization is effectively 100% because the only circuits fabricated on the chip are those required by the design.

It may be difficult to get a high ultilization in gate arrays because the auto-routing software is not always able to fit in all the tracks due to congestion in the routing channels. This is less of a problem in standard cell design where the channels can be opened up wherever necessary. A gate array chip may therefore be considerably larger than a standard-cell implementation of the design made by the same technology. The extra size increases the direct cost of fabricating gate arrays in large numbers.

Gate arrays are manufactured by almost every semiconductor company. They are available using all the circuit technologies described in Chapters 4 and 5 (CMOS, STL, ECL, CML, and others), although CMOS is by far the most popular at present. They may have up to 100 000 gate equivalents, although 10 000 is the maximum in many families and the majority of designs use only a few thousand. In choosing the type of gate array to use, the CAD capabilities and the cell library are at least as important as the design of the gate array itself, and all three aspects should be regarded as part of a single package supplied by the manufacturer.

In many of the more advanced gate array designs the chip area inside the pad ring is completely covered by transistor elements designed so that interconnections can be placed on top of the cells, thus eliminating the need for routing channels and giving a far more compact layout. These **sea of gates** or **channel-less** arrays therefore enable designs to include array circuits such as PLAs, ROM, and RAM made out of the basic elements. The electrical performance and layout efficiency are therefore greatly improved, making channel-less arrays very competitive with the best standard-cell designs but without the need for a full mask set. Some advanced gate arrays are also available with embedded, predesigned, and therefore fixed-size, ROM or RAM blocks, and even microprocessors for inclusion in chips that are otherwise designed for a specific application.

Again, 'sea of gates' is not an accurate term. 'Sea of transistors' is far better, but one has to use the more conventional notation.

At all but the lowest levels of complexity, gate arrays are usually the most economical way of implementing ASIC designs. Although their speed is somewhat less than can be obtained by optimized electrical design, it is nowadays quite adequate for the majority of applications. The chip-area penalty is also unimportant unless very large numbers of chips are to be produced and it is often more than compensated for by the low cost and the certainty of the design process.

We have seen that the elimination of many of the masks needed for standard cell design reduces the relative cost of gate arrays when small numbers are to be produced. The cost of producing even smaller, prototype quantities can be reduced even further by using masks containing up to eight designs each. Such masks are usually made by an electron beam writing on to the resist-coated mask plate in a machine controlled by data from the CAD computer. These **multi-project masks** can reduce the mask cost very substantially.

Another way of reducing commitment costs is to eliminate the need for even a single mask by using an electron beam to write directly on to the resist-coated silicon wafer for the patterning of the interconnecting layers. This process is more commonly used for the commitment of gate arrays than for the full fabrication of standard-cell ICs because they require far less time on an expensive electron-beam machine.

Gate arrays can be called 'masked' even if they are configured by direct-write electron beam in practice.

Programmable Logic Devices

Programmable logic devices (PLDs) or **user-configurable ICs** (USICs) are standard-product ICs that are electrically programmed by the user to implement particular logic functions. They are included here because modern PLDs are sometimes a viable alternative to the masked ASICs already described. Also the

Some might claim that PLDs are not actually 'designed' by the user. However, the same can be said about gate arrays where the ASIC designer is concerned with the connections in the top one or two layers only.

design methods can be similar to those used for ASICs, the only difference being in how the design is implemented on the silicon.

To program a PLD for a particular application the chip is connected to the computer used for design through a special hardware unit called a **programmer**. The final design of the PLD is downloaded to the programmer which generates a sequence of pulses on the PLD pins that configure it to produce the required digital function. In some types, programming the PLD blows minute fuses fabricated on the upper layers of the chip. In others it breaks down an insulating film between two layers of interconnect to form a conducting via, or **anti-fuse**. In both these cases the PLD is permanently changed by programming. Alternatively, in **erasable PLDs** (EPLDs) the programming information is stored in a built-in memory cell of some type so that the chip can be re-programmed if necessary. All these ICs are also called field-programmable logic devices (FPLDs). If required in very large numbers, a programmed FPLD can later be manufactured with a final mask to hardwire the function and reduce the cost.

There are many different types of PLD. They all contain the equivalent of gates for implementing combinational logic functions and most of them also have bistables. Programming is achieved by using matrices of vertical and horizontal tracks on the IC for connecting up the logic functions. The programming determines whether or not there is an electrical connection at the intersections of these tracks.

Figure 6.5 shows part of the interconnection matrix as it is usually drawn in PLD data sheets. The vertical lines are connected to the signal inputs, A, B and C, through buffers which also give the inverses of the signals. An electrical connection at an intersection is shown by an X and otherwise there is no connection where the vertical and horizontal lines cross. The horizontal lines are also shown as connected to a symbol that looks like a single-input AND gate which is a convention used to indicate that the output is an AND function of the inputs as selected by the Xs.

The type of connection at each of the X points depends on the technology

The programmer must be of the right type for the particular PLD although some programmers are fairly universal. The programming information is contained in a JEDEC file which has a standardized format.

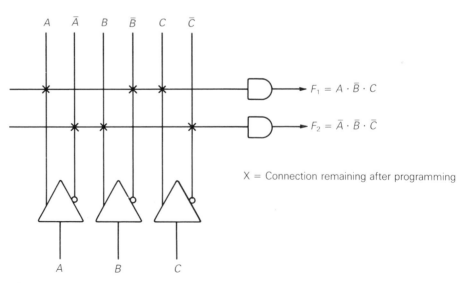

X = Connection remaining after programming

Fig. 6.5 Part of the interconnection matrix of a PLD in the form given in data books.

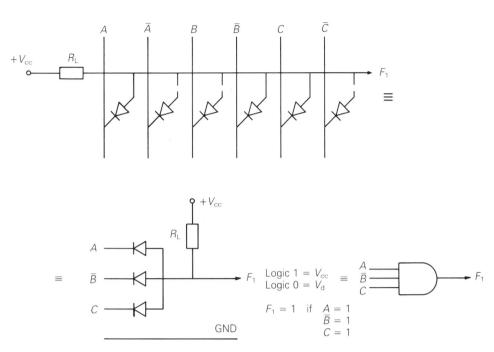

Fig. 6.6 The actual circuit of the top row of Fig. 6.5 as implemented in a bipolar fuse-programmed PLD.

used. It will be left to the reader to think about why they cannot be straight short circuits.

Bipolar Circuits for PLDs

The diode is the simplest device that can be used for making an X connection as it only passes current in one direction, avoiding the short circuits of direct connections. Diodes are used in bipolar PLDs as shown in Fig. 6.6 which is the top line of Fig. 6.5 redrawn with intact diodes at the X points. Diodes are fabricated at all the intersections on the IC, but they are disconnected by blowing the fuses in the programming process. It can be seen that the horizontal line produces the AND function, or **product term**, of the selected inputs, also commonly referred to as the **minterm**. If the outputs of several horizontal lines are connected through OR gates, any logic function can be implemented in the sum of products form. For example, combining F_1 and F_2 in Fig. 6.5 in an OR gate will give the function

See Stonham (1987) for an explanation of Boolean logic.

$$G = A \cdot \bar{B} \cdot C + A \cdot \bar{B} \cdot \bar{C}.$$

All combinational logic can therefore be implemented in arrays of this form provided that they have sufficient lines in both directions. PLDs usually contain logic blocks with 16 or more inputs and outputs.

CMOS Circuits for PLDs

Fuses are not normally used for programming CMOS PLDs because there is a better MOS alternative in the form of special memory transistors that can be used. With these the programmed connections can be erased and the chip can be re-used many times.

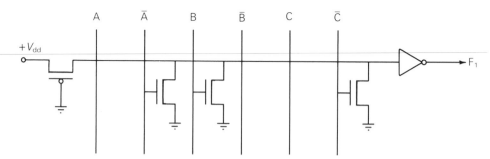

Fig. 6.7 The actual circuit of the top row of Fig. 6.5 as implemented in CMOS. The transistors are one of the stacked-gate types shown in Fig. 6.8. Only the active ones are shown.

The top line of a CMOS array producing the same logic function F_1 as previously, is shown in Fig. 6.7. This is actually a NOR circuit as the horizontal line is pulled low by a logic 1 applied to the gate of any of the n-channel MOSTs.

De Morgan's theorem states that $\overline{A} + B + \overline{C} = A \cdot \overline{B} \cdot C$.

The circuit produces the AND function by taking the NOR of the inverses of the previous inputs which is justified by DeMorgan's theorem. In the circuit of Fig. 6.7, the horizontal line is pulled up by a p-channel MOST which is permanently on because its gate is connected to ground. As this is more like an n-MOS than a CMOS gate circuit it is called **pseudo n-MOS CMOS**.

To make a programmable array using the principle of Fig. 6.7, the MOS transistors with built-in memory are made at all the intersections. These behave like ordinary n-channel MOSTs where they are required to implement the logic, but they can be programmed to be permanently off at the other locations.

The special transistors contain two gates stacked one above the other as shown in Fig. 6.8. The upper gate G_1 is just the ordinary control gate which is connected to one of the signal lines in the PLD array. The lower gate G_2 is left floating but electrons are injected into it from the silicon in the programming process. In a programmed device the static negative charge due to these electrons increases the threshold voltage from perhaps 1 to 5 volts so that it is effectively OFF for all signal voltages. An un-programmed device with no electron charge on the lower gate behaves like an ordinary MOST that can be turned on and off by the voltage on G_1.

Some manufacturers use devices with a single floating gate, which behaves almost like a diode that can be open circuited, in series with an ordinary MOST. As far as the user is concerned this makes no difference.

There are two types of memory transistor that use stacked gates in this way.

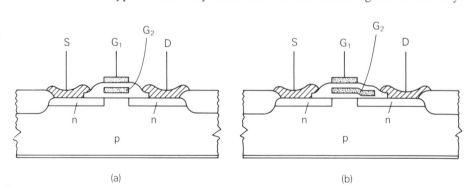

Fig. 6.8 Stacked-gate transistor types: (a) a u.v. erasable, EPROM transistor, and (b) an electrically erasable EEPROM transistor.

They differ in the method of charging and discharging the lower gate. The devices in CMOS PLDs are the same as those used in the two types of erasable programmable (EP) ROMs.

EPROMs
The floating gate is charged by applying a reverse voltage pulse to the drain that is large enough for a few of the electrons carrying the current to acquire high energies. Some of these electrons pass through the oxide into the lower gate where they are almost permanently trapped. Erasure is by exposing the IC to ultraviolet light which releases the electrons photoelectrically and returns the threshold voltage to the uncharged value (Fig. 6.8(a)).

EEPROMs
The oxide between the floating gate and the silicon is reduced in thickness near the drain so that electrons can pass through it by quantum mechanical tunnelling when the programming voltage is applied to the drain. Electrical erasure (the 'EE' in 'EEPROM') is done by applying a voltage of the opposite polarity which is both quicker and more convenient than using u.v. light (Fig. 6.8(b)).

The gate oxide in ordinary MOSTs is about 40 nm thick. In an EEPROM it is reduced to 8 nm over part of the drain to enable electrons to pass through it due to their wave nature.

In both types of memory transistor the programming voltage is produced on the chip and applied to the appropriate transistors as directed by the programming unit.

PLD Types and Architectures

Programmable logic devices are made in a wide range of types, sizes and technologies. Most of them contain programmable AND functions as described above, usually with fixed OR outputs which may be latched as shown in Fig. 6.9.

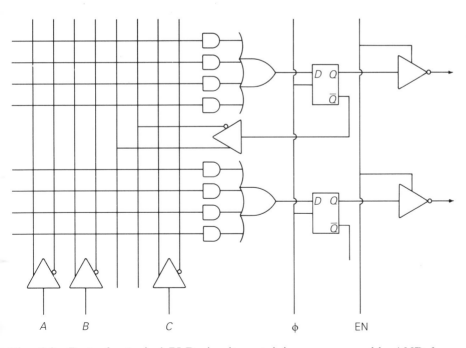

Fig. 6.9   Part of a typical PLD circuit containing programmable AND functions, and fixed OR and latch functions.

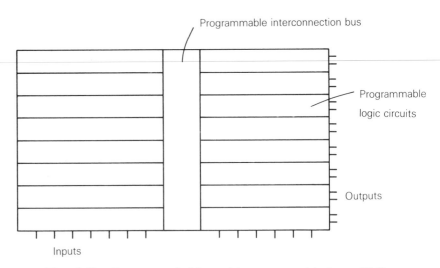

Fig. 6.10   One type of chip architecture used in large PLDs.

A PLA is the most efficient array for combinational logic because each minterm can be used several times in the programmable OR plane. With fixed OR functions the same minterm may have to be produced repeatedly for feeding into different OR gates.

A range of ICs developed by Actel uses an array of PAL-type cells that can be interconnected by programmable anti-fuses (Fig. 6.10). This makes them more flexible than standard PALs without the speed penalty inherent in the even greater flexibility of LCAs.

The Plessey **electrically reconfigurable array** (ERA) goes in the opposite direction by using only two-input NAND gates that can electrically be configured for any digital function.

These ICs can be programmed to feed the outputs back into the array to make finite-state machines for sequential logic applications. A single PLD, known as a PAL, may contain a large number of **programmable array logic** circuits of this type. In other architectures, both the AND and the OR functions are programmable which is the classic programmable logic array (PLA) structure. The familiar PROM can be regarded as a class of PLD with fixed AND connections and a programmable OR array.

PALs can produce any combinational or sequential logic function as limited only by the number of rows, columns, and OR gates in programmable circuits such as Fig. 6.9, and the number of such circuits on a chip. However, the range of applications for PALs is constrained by the inflexible architecture of rows of similar circuits. Increased flexibility is achieved in chip architectures such as Fig. 6.10 which include programmable connections between the programmable circuits. PLDs are frequently used to replace a complete printed circuit board of standard ICs because they are far more efficient as the logic is designed for the particular application.

Manufacturers of PLDs always provide CAD tools for design. As many PLDs are comparatively simple, the software is usually run on an IBM PC or compatible computer. The designer can often express the function required in a series of Boolean expressions which are converted internally into a file that is downloaded to the programmer. Some other PLD CAD systems use schematic entry of logic cells, in much the same way as will be described in Chapter 8 for masked ASICs, and again they are converted into the programming file with very little intervention by the designer.

Reconfigurable Gate Arrays

Reconfigurable gate arrays are a class of electrically programmed ICs that extend the PLD principle into a more flexible chip architecture that resembles a gate array from the point of view of the user. Reconfigurable gate arrays were first introduced by Xilinx, Inc. who call them **logic cell arrays**, or LCAs, and we

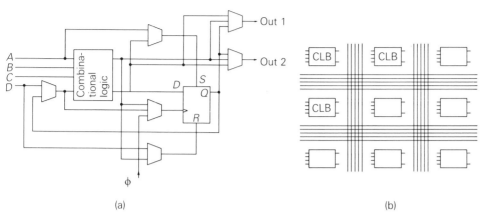

(a)   (b)

Fig. 6.11   (a) The CLB cell used in the smaller Xilinx LCAs. (b) The arrangement of cells and interconnections in an LCA. (Courtesy of Xilinx, Inc.)

will use this notation here. Advanced PLDs with a range of alternative architectures are, however, rapidly being developed by other companies.
    LCAs differ from PLDs in two ways:

1. in using a more flexible architecture, and
2. in using static RAM cells for storing the configuration data on the chip.

    The logic cells used in LCAs, called **configurable logic blocks** (CLBs), by
Xilinx, are generally similar to those in some CMOS PLDs. They contain programmable combinational logic, one or more bistables, and programmable
feedback. Figure 6.11(a) shows the CLB used in the smaller LCAs. Its function
is determined by the programming of the combinational logic and the multiplexers that permanently connect one of their inputs to the output as determined
by the programming data. The cell can be configured to produce any combinational function of the four inputs, or it can simply be a bistable, or any
combination of the two. The larger LCAs use cells with five-input combinational
logic, two bistables, and an even wider range of connections through multiplexers.
    The greater flexibility of the LCA architecture compared with PLDs is due to
the way in which the logic cells are connected. In PLDs, circuit elements such as
those in Fig. 6.9, perform logic directly on the chip input signals and they
produce direct outputs albeit with the possibility of feedback. In LCAs, however,
the cells are arranged in an array with vertical and horizontal interconnection
paths between them as shown in Fig. 6.11(b). The inputs and outputs of each
CLB can be connected to any of the interconnections as determined by the chip
configuration data. Furthermore the switching matrices at the intersections can
be programmed to interconnect any of the lines so that any CLB input or output
can be connected to any other. All the connections are made through MOS pass
transistors that are switched on or off depending on the data in the on-chip RAM
cells. The circuitry for doing this is obviously rather complex and the pass transistors reduce the chip speed.

For details see the Xilinx
*Programmable Gate Array Data
Book*, for example.

    In programming an LCA chip, the configuration data is transferred directly
from the design computer through a high-speed serial link to the RAM cells on
the chip. This can be done far more rapidly than for PLDs because it does not
need the high-energy pulses used for blowing the fuses or changing the ROM

cells of a PLD, and a programming unit is no longer required. However, the use of RAM rather than ROM cells also means that the chip programming data is lost every time the power supply to an LCA is turned off. Instead of reprogramming the LCA from the computer each time, the configuration data is therefore usually transferred first into a special electrically programmable ROM chip via a simple programming unit. The ROM is then mounted on the printed circuit board adjacent to the LCA and the configuration data is automatically loaded into the LCA every time the system is powered up.

LCAs are made in a range of sizes, currently with up to 320 CLBs, equivalent to 9000 gates, per chip. This makes them comparable to the most commonly used gate arrays with which they have many similarities. The main difference is that the basic cells are necessarily more complex and the interconnections are determined by down-loading the design data rather than fabricating the masks for committing a gate array. The benefit of this is that LCAs can be configured by the designer without the cost and delay in getting chips fabricated by a semiconductor company. Also the design can be altered almost instantly either in development or for a revised product.

The flexibility in the architecture of LCAs can only be used effectively with the appropriate CAD tools. The design system is almost the same as that used for any cell-based technology including gate arrays and some of the more advanced PLDs. The required circuit functions are decomposed into interconnected library cells or **macros** and entered into the computer for simulation and checking. When the design is correct, the software can determine (a) which CLBs are to be used for each macro, (b) the configuration information for each CLB, and (c) the interconnection routes which are also specified in the programming data. The resulting **layout** can also be viewed and edited by hand if necessary. It is not surprising that a fairly powerful personal computer is required to do all this.

All electrically reconfigurable ICs have one inherent advantage over other chip technologies. It is that the chips can be 100% tested in production by configuring them in such a way that externally applied test programs can reach every part of the circuit. The configuration data can then be erased before they are reprogrammed for use. In all other technologies it is extremely difficult to achieve full testing from the external connections of a dedicated chip function as we will find in the next chapter.

The cells on LCAs are far more complicated than the basic elements used in gate arrays because the number of interconnections is limited when they are electrically programmable. More functionality therefore has to be built into the individual cells instead of using combinations of simple cells.

Comparison of ASIC and PLD Technologies

A digital designer needs a good understanding of the special chip technologies available for producing digital ICs in small numbers in order to decide on the best one to use. The four technologies described in this chapter are standard cells, gate arrays, PLDs and LCAs. PLDs can only be used for the smaller digital functions, currently up to a few thousand gate equivalents, although their architecture is often thought to be a limitation to their use even below this. LCAs can be used for similar applications but above about 10 000 gate equivalents the choice is narrowed to standard cells and gate arrays (and of course full-custom) if a digital function of this size must be implemented as a single chip of silicon.

In comparing the technologies for the range of complexities where they overlap we should ideally consider chips produced by identical fabrication processes. In practice this is not possible because commercial products are seldom made in exactly the same ways. We can, however, find general trends enabling them to be compared in terms of performance and cost.

Performance

Performance is measured by the maximum chip operating frequency and the power consumption. The highest speed is always achieved in full-custom design where the individual circuits are optimized to use the fabrication technology to the best possible advantage. Standard-cell ICs lose some of this speed because the circuits are not exactly tailored to the particular application and the layout increases the interconnection capacitances. Gate arrays are slightly worse again because the circuit cells are all built from identical transistors. It is difficult to obtain precise figures for the fall-off in speed since it depends on the particular design but we might guess an average reduction of 10% for standard cells and a further 5% for gate arrays compared with full-custom design. The power consumption might increase by similar amounts but in CMOS it is compensated for by the lower operating frequency.

PLDs and LCAs are slower than any of the masked ASIC technologies made by the same fabrication processes because of the extra interconnect delays inherent in electrically programmable architectures. The extra lengths of the connections increase the capacitive loading on gates and the pass transistors used in the connections on LCAs add resistance which introduces extra *RC* time constants. However, advances in fabrication technology have improved the speed of programmable ICs to such an extent that many of them are now more than adequate for most applications even though they fall short of the ultimate speed of full-custom circuits.

Cost

The cost of obtaining small numbers of non-standard ICs is usually of far greater importance than their performance. In comparing technologies it is essential to include the costs of design and, where necessary, masks, in addition to the direct cost of the chips. Because of these fixed costs, the cost per chip depends greatly on the number produced. Costs are always very variable, depending on commercial as well as technical factors, and up-to-date estimates should always be obtained before starting on a particular design. The cost trends for the different approaches are, however, as shown in Fig. 6.12.

At the low-volume end, the cost is dominated by the non-recurrent engineering (NRE) charges which are the design and mask costs. At the high-volume end, the NRE charges are amortized over the large number produced and the chip cost is dominated by the costs of silicon manufacturing and packaging.

The cost of full-custom ICs is seen to be greatest except when they are produced in very large numbers indeed. This is because of the very long time required to design them and the likelihood of needing two or more full mask sets before the design is correct. On the other hand, a full-custom chip has the most compact layout allowed by the technology, giving the smallest chip size for a given function and making it inherently cheaper in mass production.

Semi-custom design methods reduce the design time, perhaps by as much as ten times, and increase the design reliability. A full mask set is required for standard-cell designs but the number of masks is reduced for gate arrays which are therefore initially cheaper. However, a standard cell chip is generally smaller

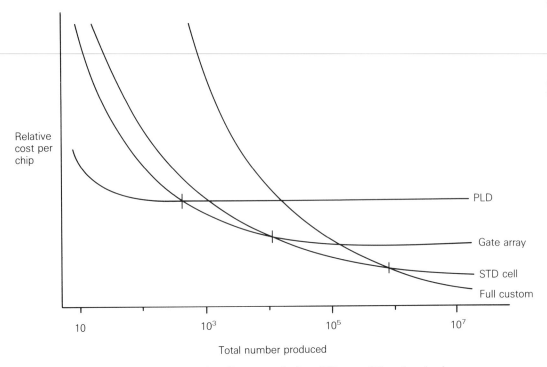

Fig. 6.12   Cost trends for different IC technologies.

because of the low utilization of the silicon in most gate array designs. Standard cell ICs are therefore eventually cheaper to produce in large numbers.

Programmable ICs eliminate the need for masks which is a great advantage for very small numbers even though the cost of the chips themselves is generally higher than for masked ASICs. This is partly because a programmable chip is considerably larger than a gate array with the same utilization due to the space taken by the programmable architecture. The most important advantage of programmable ICs is, however, that a design can be tested immediately it is completed without any delay for chip fabrication so that they are ideal for prototype development. Depending on the CAD system used it may be possible to transfer the design to a masked ASIC subsequently with very little further effort.

No reliance should be placed on the cross-over points for the curves in Fig. 6.12. As the curves are so shallow, small cost changes can make a large difference to the production volume at which one technology becomes preferable to another.

Summary

This chapter has described the special chip architectures used for ASICs where the number of chips to be produced for any design is relatively small. For ASICs to be used economically, the design cost must be reduced as far as possible by using semi-custom methods based on libraries of basic cell types. The design process determines the positions of the cells, assembled in rows or larger blocks, and the connections between them to implement the required chip function. In standard-cell ICs the circuit detail, which is added for chip fabrication, is contained in all mask layers. The cost of producing a small number of ICs is, however, greatly reduced by using gate-array chips which have large numbers of identical components in the lower levels committed to the particular chip function by one or two masked layers of interconnect. The silicon area is used less efficiently in gate arrays but the loss is reduced in the channel-less architecture.

Electrically programmable ICs contain logic blocks and interconnections that can be configured by downloading a data file from the CAD system. Early PAL architectures were inflexible for general-purpose use but modern reconfigurable arrays are comparable to gate arrays for designing digital systems with up to 10 000 gates and they have similar performance.

Problem

Calculate the cost per chip of producing quantities of between 100 and 100 000 ASICs made by gate-array, standard-cell, and full-custom technologies using the data given below. Plot the results as in Fig. 6.12 using log–log scales and determine the production volumes over which each technology has the lowest cost.

$$\text{Cost/chip} = \frac{mD + nM}{N} + \frac{W}{Y} + P,$$

where: N is the number of chips produced; m is the design time (man months); n is the number of custom masks required; Y is the wafer production yield; D is the design cost per man month (take as £3000); M is the cost per mask (take as £1200); W is the processing cost per wafer (take as £400); and P is the test and packaging cost per chip (take as £2.50).

The data for each technology is given in Table 6.1. In each case the chips are to be made on wafers of 100 mm diameter and the yield is 40%.

(This Problem was provided by Professor P.J. Hicks, Department of Electronics and Electrical Engineering, UMIST.)

Table 6.1

	Gate array	Standard cell	Full custom
Number of custom masks	1	8	8
Chip size (mm)	5.0×5.5	4.5×4.5	3.5×4.0
Design time (man months)	3	3	18

7 Digital Systems Design for ICs

Objectives Studying this chapter should enable you
☐ To start to design digital ICs at the system and block-diagram levels.
☐ To understand the differences between functional and hardware descriptions of digital systems.
☐ To see the important role of simulation in IC design and the features of different types of digital simulator.
☐ To learn about the importance of accurate timing in synchronous digital systems and ways of achieving it by careful design at the gate level.
☐ To understand the problem of testing complex ICs and some of the methods used to ensure that an IC is fully testable.

The subject of digital electronics is concerned largely with the design of logic circuits made up of gates and bistables. It is based on Boolean algebra and it uses techniques such as Karnaugh maps for combinational logic and state methods for sequential logic design. These techniques are widely used in the design of digital ICs and we will assume that the reader is already familiar with them as they are normally covered in any course in elementary electronics.

There are many books on digital logic design, a few of which are listed in the Bibliography.

ICs are certainly digital circuits but they are far more complicated than the circuits described in textbooks on digital electronics even if they contain as few as 1000 gates. To avoid being overwhelmed by the detail of a digital IC it is useful to think of it as a digital system rather than a circuit. We saw in Chapter 1 that the problems in designing a very complex system can be reduced by working down a hierarchy of levels, starting from the overall block diagram and splitting it into sub-blocks, perhaps several times, until the recognizable digital functions such as registers, counters and adders are reached. Most of the creative work in designing complex ICs is in fact concerned with blocks of circuitry rather than gates or bistables, and the techniques for this level of design are often far less familiar than for low-level digital design.

In this chapter we are concerned with some aspects of digital systems design that are used in designing ICs. These are:

1. The use of simulation,
2. The need for precise system timing, and
3. The problem of testing complex systems.

System Design and Simulation

The formal design of systems at the block diagram level starts by developing algorithms for producing the required behaviour and simulating them using an appropriate high-level computer language. When the algorithms are shown to work correctly, they are mapped into a top-level block diagram for the system which is then successively partitioned into more and more detailed block diagrams.

Partitioning is the splitting up of a block diagram into smaller blocks. It may not be easy to determine the optimal partitioning of a complex function.

108

For simple ICs the top-level design at the block-diagram level is often done intuitively rather than by using formal methods. The chip specification may immediately suggest some obvious divisions of the system into a number of interacting blocks such as logic units, memory, and input and output interfaces, and how they might be connected. The block diagram is developed by considering the function of each block of circuitry, that is how the output data is related to the input and how it is affected by the control signals. Such an approach can be used for Worked Example 7.1.

The algorithmic state machine (ASM) method provides another more formal method for designing digital systems. See Green (1985).

Worked example 7.1

An integrated circuit is required to operate an electronic door-lock. If the six-bit signal from a bank of push buttons is not the right combination to open the lock the circuit is to be disabled for five minutes during which time it will not even respond to the correct input. Suggest a block diagram for the IC.

Solution A first idea for the solution is given in the figure below. It may or may not function correctly.

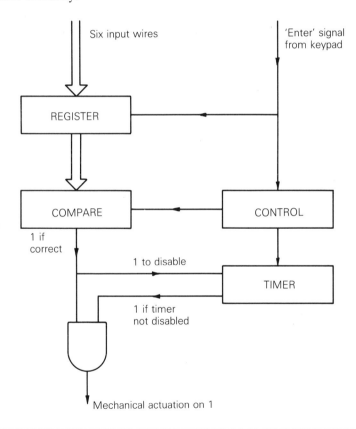

Exercise 7.1

Improve the block diagram of the electronic lock IC proposed above. You will need to make additional decisions about its exact function and to implement them in specifying the blocks. Make sure that it will not be insecure (a) in the intervals between entering new combinations, and (b) if the input buttons are pressed extremely slowly.
What does the control unit do?
If the correct combination is entered, does it stay in the register for ever?
How could the combination to open the lock be changed?

Exercise 7.2   Write a full specification to describe exactly what the electronic lock IC has to do in every possible condition.

Exercise 7.3   Suggest an alternative block diagram for the electronic lock IC.

Most digital ICs can be thought about in the same way even though they may be very much more complicated than the above example. Whether they are parts of communications, computer, control, or other systems, it is nearly always possible to think in words about the overall function and the basic operations that have to be performed for it to work correctly. This can be described as the **functional** or **algorithmic** level of design, and it is entirely concerned with the relationship between the input and output signals of the main blocks of the system. The signals, their timing, and their effects on other signals must be precisely defined at this level to make sure that the system operates correctly in all eventualities. Functional design therefore considers exactly what each block does but not how it does it and, for this reason, it is sometimes called **behavioural** design.

Functional design can be carried through any number of levels, dividing up the main blocks into smaller blocks at each stage and always with the functions and signals being fully defined.

Exercise 7.4   Suggest suitable sub-blocks for the control and timer units of Exercise 7.1. Write down the function of each sub-block and define the signals that flow between them. The external signals for each unit should be the same as in the previous exercise.

The splitting up of the blocks is not done entirely from functional considerations. The designer doing the job always has in mind the digital hardware that might eventually be used to build the system. A very wide range of functions can be included in ICs, for example data registers with serial or parallel inputs and outputs and many types of control, counters with similar features, arithmetic units, memory blocks, combinational logic blocks, and state machines for producing control functions. For instance, in considering the timer unit in Example 7.1 the designer should know that some type of oscillator will be needed with a counter to divide the frequency down for timing long periods. The functional decomposition of the block diagrams continues until the identifiable functions of digital hardware such as these are reached. A list of such functions is given in Appendix 5.

The conversion of functional description into hardware descriptions of circuit blocks is called the **register level** of the design process and it is the most important step in changing abstract ideas into physical circuitry. Figure 7.1 shows by means of an example how the relationship between signals is changed into a relationship between real circuit blocks connected by the physical equivalent of wires. Once this is done the design of the blocks can proceed by the ordinary methods for designing digital circuits using various types of gates and bistables.

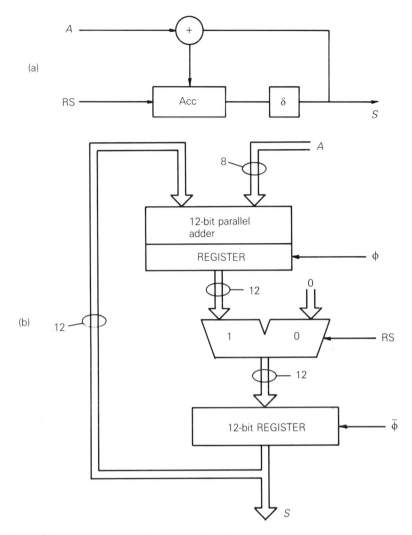

Fig. 7.1 (a) A functional diagram of a digital system. *A* is a number in the range 0–255 that is constant for a period δ. *S* is a number in the range 0–4096. Add *A* to the contents of the accumulator and output after a delay δ. Set ACC to 0 if RS ⇒ 0. (b) A hardware diagram of the same system the adder output is latched and written into the accumulator when φ ⇒ 1. *A* must be steady during φ = 0. RS must be steady during φ = 1.

Hardware Description Languages

A 'hardware description language', or HDL, is a high-level computer language used for writing descriptions of electronic system blocks. It can be used to describe either the behaviour or the structure of the hardware. The 'behaviour' refers to the relationship between the input and output data, and the 'structure' refers to the sub-blocks, registers, gates, etc., and their interconnections. The highest level of a system description, the specification, tends to be fairly abstract, giving information such as algorithmic procedures relating to the overall be- haviour. In the design process 'behaviour' is translated into more and more detail on 'structure'. Many HDLs can be used for both types of description and they incorporate hierarchies for developing functional blocks right down to gate-level descriptions. Some HDLs also include procedures for 'logic synthesis' which is the automatic generation of gate-level logic from a behavioural description such as a Boolean expression.

HDLs can be general-purpose, procedure-based computer languages such as Pascal, or they can be written specifically for CAD purposes. They contain the features of programming languages such as loops and conditional statements, and their hierarchical structure enables the design to be expanded in a systematic way. One of the most powerful and flexible hardware description languages is VHDL which is likely to become a widely used standard in the future.

Details of the features of HDLs and their notation can be found in the manuals provided with the software.

Simulation

The design of a digital system should always be checked at every stage using computer simulation. A simulator program uses:

1. A description of the function of the circuit elements or blocks,
2. A list of inputs, outputs and interconnections, and
3. Input waveforms, which for digital systems are sequences of 1s and 0s.

The program then calculates the output waveforms. If the designer finds that they are not as intended, the interconnections or the functions of the blocks can be altered and the circuit re-simulated repeatedly until it is correct.

There are many types of simulator used for IC design. They differ primarily in the ways in which the circuit elements or blocks are described or **modelled**. They can be divided into:

1. Logic simulators,
2. Timing simulators, and
3. Functional simulators.

A simple logic simulator might be used for checking the operation of a gate circuit as in Fig. 7.2. A simulator program of this type already contains models of the NAND and NOR gates and the designer supplies it with the list of connections and input waveforms in the appropriate format. Running the program then produces the waveform for the output F, either graphically or as a table of values. In this particular case, simulation is scarcely necessary because the logic function of the circuit is so easily checked by hand. However, the value of simulation, even at this level, is to make certain that the designer has not made any mistakes, and this is done by exercising the circuit fully with an appropriate test waveform. For larger circuits simulation is obviously far quicker than hand methods for checking the operation.

A more realistic simulation would include the effects of input–output delays in the gate models, and this would show the presence of unwanted transient voltages, or 'glitches', at the output due to different path delays in a circuit such as in Fig. 7.2. The simulation of a larger circuit containing a part that generates glitches will also show whether the unwanted signals give errors elsewhere in the system. If they do, the design will have to be changed to eliminate the possibility of glitches occurring.

Some examples of the features of commonly used simulators will be given in

Examples of HDLs include HELIX, and ELLA that will be mentioned again in Chapter 8, and the MODEL language which is part of the ES2 CAD system.

VHDL stands for Versatile Hardware Description Language. It was produced as part of the very high-speed integrated circuit (VHSIC) programme in the USA.

Simulation result for the circuit of Fig. 7.2 but including gate delays. The glitch in *F* is not logically correct. It occurs with the signals shown because of the slightly different times at which the edges reach the NOR gate from *A* and from the output of the NAND gate.

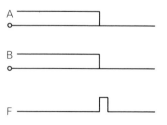

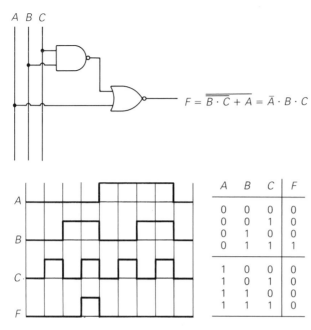

$$F = \overline{\overline{B \cdot C} + A} = \bar{A} \cdot B \cdot C$$

A	B	C	F
0	0	0	0
0	0	1	0
0	1	0	0
0	1	1	1
1	0	0	0
1	0	1	0
1	1	0	0
1	1	1	0

Fig. 7.2   A gate circuit and its simulation results.

Chapter 8. Many of them support hierarchical descriptions of circuits so that a complicated gate-level circuit can be regarded as a single unit for connection to other circuits, enabling larger and larger blocks to be built up. Some of them also contain easier ways for describing multiple inputs on parallel wires, for example by replacing sequences of 1s and 0s with decimal or hexadecimal numbers.

Simple ICs can usually be designed using gate-level logic simulation only. However, the full simulation of a complete chip with a comprehensive test waveform and properly modelled gate delays requires quite a lot of computer power and it may take a long time to run on a small machine.

For large ICs, a functional or behavioural simulation is far more efficient because it enables the operation of an IC to be checked quickly at the block diagram level. A simulator of this type uses functional models for the circuit blocks written by the designer in an appropriate hardware description language. Functional simulation is extremely important for the system levels of design because it can add rigour to what would otherwise be an intuitive process. In developing the block diagram in detail it is therefore re-simulated at each stage to verify that it still functions correctly, and this continues down to the register level where models based on gates and bistables can take over. **Multi-level simulators** can work with parts of a system described by behavioural models and other parts by gate-level models.

Simulator Use

The first step in using a simulator is to enter a description of the connections between the circuit blocks or elements into the computer. This is done either graphically, in **schematic entry**, or by typing in a text file (**netlist entry**). In schematic entry, the circuit is drawn on the display screen, usually by calling up symbols for the circuit elements from a menu, putting them in the right positions

'Schematic entry' is sometimes called 'schematic capture'.

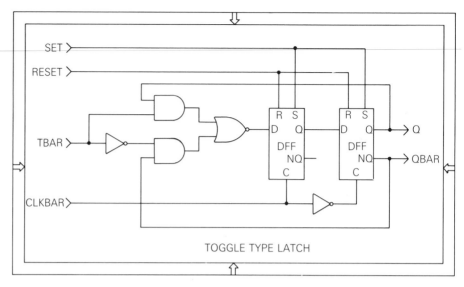

Fig. 7.3 An example of schematic entry of a circuit containing both gate level and D-type symbols as it appears on the computer screen.

The netlist for this circuit might look something like CCT TTYPE1 (SET, RESET, CLKBAR, TBAR, QBAR, Q)
Element Declarations
NOR
    NOR1 (NET4, NET6, NET5);
INV
    INV1 (NET1, TBAR);
    INV2 (NET2, CLKBAR);
DFF
    DFF1 (Q, QBAR, NET2, NET3, SET, RESET);
DFF2 (NET3, CLKBAR, NET4, SET, RESET);
AND
    AND1 (NET5, NET1, QBAR);
    AND2 (NET6, NET1, QBAR);
Wire Declarations
INPUT TBAR CLKBAR RESET SET;
WIRE QBAR;
UNID NET6 NET5 NET2 NET1 NET4.

The element and wire names are arbitrary and the format depends on the particular simulator used. This is HILO.

The lowest level gate description of an IC contains no hierarchy and is said to be 'flat'. Designs are often 'flattened' for final simulation, which may take a great deal of computer time because of the large amount of detail in a gate-level description.

using a mouse or other pointing device, and adding lines representing the connecting wires. Some symbols, such as those for gates, etc., will already be stored in the system's symbol and cell library. Others are created by the designer in the form of rectangular boxes with lines representing the input and output wires of blocks of circuitry. The behavioural descriptions of blocks created by the designer, written in a suitable HDL, must, of course, be added to the system library.

Schematic entry is more than just a computer drawing of the circuit as it also provides the data for simulation. The final drawing is therefore usually labelled with simple alphanumeric names for the symbols and for the input and output connections (or **pins**), and these names are used in all later references to the circuit. Some simulators also require the addition of **attributes** to the symbols. These are the numerical values of the delay parameters used in simulating the circuit represented by the particular symbol. An example of the schematic entry of a simple circuit is given in Fig. 7.3.

The information contained in a schematic circuit diagram is compiled by the computer, first into a text form and then into machine-readable code. The text form, or **netlist**, is a list of the connections between the named pins and components written in a format that depends on the system used.

As an alternative to schematic entry, the circuit information can be typed directly into the computer using either a behavioural or a structural HDL description. The textual description is very concise at the higher levels but its length increases on working down the hierarchy and a gate-level netlist may involve a great deal of typing. Many systems, however, allow very concise gate-level descriptions of blocks containing repeating cells such as multi-bit registers or counters, allowing them to be typed in one or two lines of text. Details will not be given here because the notation depends greatly on the particular CAD system used but they can be found in the manuals provided with the software.

Further information, such as the functional models of designer-created symbols, input waveforms, and control statements also have to be typed into the computer

114

before the simulation can be run. The output can be in the form of waveforms or tables giving the logic values at the required points.

We will consider the features of some particular simulator packages in Chapter 8.

System and Circuit Timing

Synchronous Systems

In any dynamic system, careful attention has to be given to the timing, particularly when there are many things happening simultaneously as in an IC. Many digital systems, including all but the simplest ICs, are synchronous, that is the timing is determined by a continuous clock signal that triggers every single event. A **single-phase clock** is usually a square-wave with a period T, and frequency $f_\phi = 1/T$ Hz.

A single-phase clock waveform.

The advantage of synchronous timing is that nearly all the signal changes, $0 \rightarrow 1$ or $1 \rightarrow 0$, start at times close to one or other of the clock edges. Each circuit takes a certain time to respond and in Chapter 1 we defined the propagation delay, and the rise and fall times. All these times depend on many factors such as gate loading (predominantly determined by the fan-out), manufacturing tolerances in the β and V_T values of the transistors, temperature, power supply variations, etc., and precise values cannot be relied upon. In a synchronous system, the clock period is long enough for even the slowest circuit to respond fully before another change occurs. The faster circuits therefore have to wait for the slow ones to settle before the next clock edge allows the data to propagate. Synchronous design is reliable because the signals can always be predicted and errors due to variable delays are eliminated by design.

Since the chip speed depends on the slowest events that take place in half a clock cycle in a synchronous system, it is important to identify, and perhaps eliminate, the longest delays in the design process. Long delays occur if gates are connected in series for multi-level logic functions so that wherever possible there should be only one level of logic evaluated in each half clock cycle. The number of levels can often be reduced by using different logic implementations or by splitting the evaluation between gates operating on alternate half clock cycles, as in a pipeline. Long delays are also caused by gates with a high fan-out which can also be avoided in the logic design. A more fundamental source of long delays, however, occurs in circuits such as adders or comparators where a carry signal is passed from one circuit to another. Many ingenious methods have been devised for reducing this problem in the design of such circuits.

See, for example, Weste and Eshraghian (1985).

Edge-triggered Bistables

Synchronous systems operate by latching the signals into bistables at the instants when the clock changes from 0 to 1 or, less commonly, from 1 to 0. A positive edge-triggered D-type bistable (Fig. 7.4) takes in the signal on its input D when the clock goes from 0 to 1, and stores it, giving steady outputs, Q and $\bar{Q}$, until the next positive-going clock edge when the cycle repeats.

The 'D' in 'D-type' stands for delay. The D-type is the simplest data storage circuit and it is by far the most commonly used type of bistable in ICs.

In practice, both the clock and the data signals applied to a bistable are rounded off as shown in Fig. 7.5. To ensure that the latching-in of the data is completely reliable, the input signal needs to be stable for a short time before the clock edge. The minimum time between the application of the input and the clock is called the **set-up time** t_{su}, as defined in Fig. 7.5. The input also has to

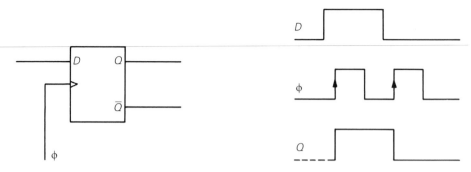

Fig. 7.4   The symbol and function of a positive edge-triggered bistable.

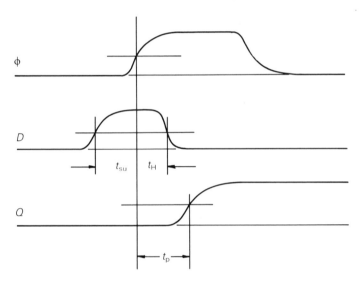

Fig. 7.5   The set-up and hold times t_{su} and t_h of an edge-triggered D-type. t_p is the propagation delay

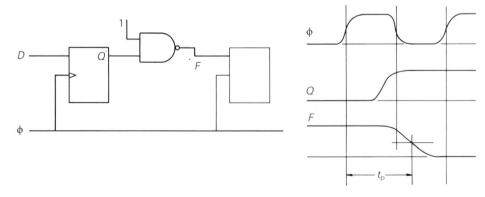

Fig. 7.6   Waveforms in a typical synchronous circuit element driven at near the maximum clock frequency.

remain constant after the clock edge for a time greater than the **hold time** t_h to be certain of it latching-in. These times are given for D-types in cell libraries or they can be calculated for full-custom circuits.

The timing of a circuit containing a positive edge-triggered D-type driving a gate at near its maximum speed is shown in Fig. 7.6. During the time when $\phi = 1$, the latch is responding to the input $D = 1$, and the gate is starting to change its state. During $\phi = 0$, Q stabilizes and the gate responds completely so that the output F is steady by the time of the next positive going clock edge. The total delay time t_p between the input reaching logic 1 and the output of the gate becoming true is nearly one clock period at the maximum clock frequency.

In a real circuit the combinational logic in between D-types would often be far more complicated than the single gate shown in Fig. 7.6, so that the delays would be greater and the clock frequency would have to be reduced. This is why it is so important to ensure that none of the combinational logic delays become excessively large. If multi-level logic is used, variable delays through different paths can lead to spurious transient voltages at gate outputs. These glitches have no effect if they die away before the set-up period of an edge-triggered bistable connected to the output.

Two-phase Clocking

In MOS circuits, the D-types are often made using switches because the circuits are far simpler than those built of gates. A transparent latch using switches is shown in Fig. 4.21. To make an edge-triggered D-type using the master–slave principle, two transparent D-types are connected in series, the first clocked by ϕ and the second by $\bar{\phi}$, as described in Appendix 4. For reliable operation the switches controlled by ϕ and $\bar{\phi}$ must open and close at *exactly* the same times because the output would be uncertain if they were both closed together. The edge-triggered D-type designs supplied in cell libraries are guaranteed to work correctly and they can be used safely in semi-custom ICs. However in full-custom designs, the designer cannot be as confident of correct operation without a great deal of practical verification and an alternative is to replace the single clock signal by two separate clocks of the same frequency but of different phases.

Figure 7.7(a) shows the waveforms ϕ_1 and ϕ_2 of such a two-phase non-overlapping clock scheme. One phase is used for the master section of a master–slave bistable and the other for the slave, with the gap t_{12} eliminating any possibility of both of them being on together (as shown in Fig. 7.7(b)).

A two-phase non-overlapping clock never has $\phi_1 = 1$ and $\phi_2 = 1$ at the same time.

This two-phase clocking scheme is commonly used in full-custom design, not just because it gives reliable bistables but also because it divides every clock period T into four parts, during which different circuit operations can take place. This gives greater flexibility in full-custom circuit design, as will be seen in Chapter 10. Two-phase clocks are not normally used in semi-custom ICs, however, because they have no advantages with standard circuits and the problem of generating the second phase on the chip is avoided.

Asynchronous Inputs

In a synchronous system, the timing of all the transitions is controlled by the clock and this avoids problems that can arise due to variable circuit delays. In asynchronous systems, on the other hand, the transitions depend solely on the timing of the input signals and special design techniques are required to ensure

See, for example, Stonham (1987, pp. 88–107).

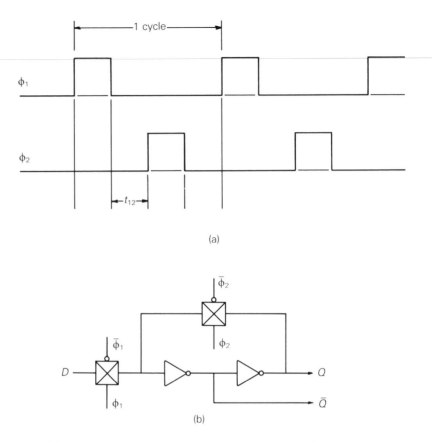

Fig. 7.7 (a) Clock waveforms in a two-phase scheme. (b) A two-phase CMOS transparent D-type circuit.

correct operation with variable propagation delays in different paths. Due to the complexity and inability to make changes to ICs, asynchronous design should be avoided for all but the simplest ASICs.

Whenever possible, integrated circuits should therefore be completely synchronous and care must be taken to see that asynchronous events are not propagated inside the IC. Input signals are not synchronized unless they come from a source running off the same clock as the IC, but otherwise they should be sampled synchronously by the first circuits they encounter on the chip. This is done by including latches for all inputs, using simple transparent D-types that are quite adequate for this application.

The problem of timing the input signals is largely the responsibility of the systems engineer using the IC, but it can be greatly reduced by using a handshaking arrangement with data-request and acknowledge signals generated on the chip. Similar arrangements may also be included in the output interface.

Handshaking is described in Downton (1984) in the context of microprocessor interfacing.

Circuit Precautions for Reliable High-speed Operation

As the frequency of the clock driving a synchronous IC increases, a point will be reached where errors start to be generated. Careful circuit design is needed to get the maximum safe frequency as high as possible, and this requires the same attention to timing at the circuit as at the logic level. Some simple precautions

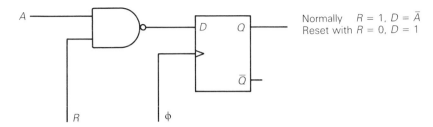

Fig. 7.8 Addition of a reset input to a D-type.

should be observed to reduce **clock skew** which is the displacement of clock edges in some of the circuits and which may cause errors elsewhere.

1. Never pass the clock signal through a gate. If a D-type needs to be turned on and off use an input enable, as described in Appendix 4, rather than turning the clock off locally.
2. Ensure that all paths distributing the clock signals are driven by a symmetrical arrangement of buffers so that all edges remain in synchronism.
3. If possible use an external clock source rather than an on-chip clock generator. This will enable the frequency to be changed which helps greatly in finding the source of errors. However, the chip specification may require the clock to be generated internally to simplify the PCB on which it is mounted.
4. Do not use asynchronous R–S bistables. If the use of a set–reset function is unavoidable, an enabled D-type bistable should be used, Appendix 4.
5. A global reset must be provided to set all the bistables to a known condition before starting to test the chip. This reset should be synchronous and it can be incorporated by adding a NAND gate to the input path of the bistables as shown in Fig. 7.8. Local clear and preset inputs can be included in any bistables requiring them, but their control signals should come from a synchronous source.
6. Never use extra inverters to generate delays in an attempt to make a circuit work correctly, because the delay values are too variable. There should be no need for delay elements if correct synchronous design techniques are used.

Chip Testing and Testability

In Chapter 3 we saw that defects are inevitable in the production of anything as complex as an IC. The object of testing is therefore to separate the good from the bad chips in production, but how to do this economically raises many problems that have to be solved by the designer.

Two distinct types of test are made on all completed wafers before separating the individual chips.

1. Process and parametric tests made on **drop-ins** which are chip areas containing structures for evaluating process parameters, threshold voltage, delays, contact resistance, etc. of concern to the semiconductor manufacturer.

   The drop-ins are easily seen on wafers as in Fig. 2.1.
2. Functional tests made on every chip to see whether it has the correct logic operation.

The IC designer is responsible for deciding exactly how functional tests will be performed.

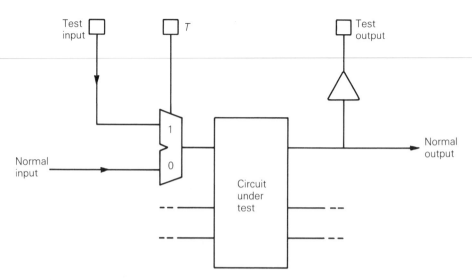

Fig. 7.9   Use of multiplexers for isolating a circuit block for testing.

Functional Testing

The problems of functional testing are due to:

1. The complexity of the function and the large number of transistors in the complete circuit, and
2. The inaccessibility of most of the circuits since test signals can only normally be applied at the external connections unlike a digital system on a printed circuit board where many internal nodes can be probed.

A production chip tester applies sequences of 1s and 0s, called test **vectors**, through probes to the input pads, examines the outputs with similar probes on the output pads, and compares them with expectation. High-speed chip testers are extremely expensive pieces of equipment and the time taken in testing each chip has to reduced as far as possible so that the cost of testing does not outweigh the fabrication cost.

Testing tries to find out whether the IC will perform correctly with every possible input signal. At first sight one might consider applying every combination of 1s and 0s to the inputs to test it fully. For a chip with N input pins the number of vectors for such an exhaustive test is 2^N. For $N = 16$, this is about 65 000 tests which could be completed in 6.5 ms on a tester working at 10 MHz, and this is quite feasible. However, most ICs contain internal latches, each of which has two states, and to cover all eventualities every combination of inputs and states would have to be tested. If the chip contains M bistables, the number of tests required is therefore 2^{M+N}. A very modest IC could easily have $M = 32$, $N = 16$ which would require 2.8×10^{14} tests, taking about 11 months at 10 MHz! This is clearly impossible and we have scarcely started to get near to the complexity of even LSI circuits, so that totally different methods have to be used.

Dramatic reductions of test time can be achieved by partitioning the IC into separate blocks for testing. Figure 7.9 shows how a multiplexer can be used to isolate the inputs of a block, allowing test signals to be applied and evaluated when the chip is in **test mode**, $T = 1$. If an arrangement like this was used to divide the previous IC into four parts, each with $M = 8$ and $N = 4$, the number of tests could be reduced to only 16 000 requiring 1.6 ms. Although this is a totally artificial example it does illustrate the value of splitting up the function for testing purposes.

Another way of shortening the test time is to apply a reduced set of test vectors, chosen for their particular ability to detect faults. Even a faulty circuit will give the correct output for many of the possible inputs, and faults will only show up with particular input combinations. We therefore have to devise a sequence of test vectors that give outputs that are changed by fabrication faults and to do this we need to know what effects faults have on digital circuits.

Fault Modelling

Although many types of physical faults may occur on an IC their effect at the digital level is often that a gate output will fail to change its state from 1 to 0 or 0 to 1 when it should, that is the output is stuck at logic 1 or logic 0. This is called the **stuck-at-fault** model and it is widely used in the design of digital systems of all types including ICs even though it may not be very accurate for faults such as breaks in tracks or inadvertent leakage paths.

Stuck-at faults may occur on any of the circuit nodes of an IC. Although a chip may have more than one fault, it is usually regarded as adequate to apply tests designed to show the presence of single faults only, on the grounds that it is most unlikely for the effects of two faults to cancel each other out and allow a doubly faulty chip to pass as good. If the IC contains n nodes, there are 2^n possible single faults (each node stuck at stuck at 1 or stuck at 0) and the designer must produce a relatively short set of test vectors that will show up the occurrence of as many as possible of them.

To show how test vectors are composed, we start by considering the two-input NAND gate with the output stuck at 1 as shown in the margin. Inputs 00, 01 and 10 will all give the correct output, namely the logic 1 even though it is fixed. The only input to show up the fault is therefore 11 which still gives $F = 1$ instead of the expected logic 0. If the output of the same gate was stuck at 0 instead of 1, the fault would show up with any of the inputs, 00, 01 or 10 but not by 11. Two test vectors are therefore adequate to show up both faults in this case. Similar reasoning can be applied to stuck-at faults at the input of a gate.

Testing a gate always involves trying to change the state of its output in this way. The problem on an IC is that the gate may be deep inside the circuit and yet it still has to be reached by test vectors applied to the inputs. Also the gate output may be a long way from a chip output connection where its effect has to be observed. The theory of test pattern generation therefore considers the **controllability** of every gate input and the **observability** of the outputs. This theory is often included in digital design courses and books so that we are not concerned with it here.

To detect the faults in some part of a circuit, the designer devises test vectors to give outputs that are different from the fault-free outputs. Any pair of input vectors may cover several of the possible faults, and these faults are said to be **covered** by these vectors. The proportion of faults covered by a complete sequence of tests is called the **fault coverage** and the designer should aim for a figure close to 100% before submitting a design to fabrication. This may be very tedious but luckily it can be done automatically for combinational logic in some CAD systems. The problem is far greater for sequential circuits.

The multiplexer is equivalent to a two-way switch controlled in this case by *T*.

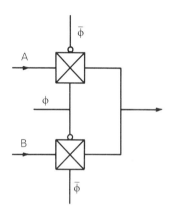

When *T* = 1 the output is connected to *A* and when *T* = 0 it is connected to *B*. In CMOS circuits the multiplexer is made using two transmission gates.

The stuck-at fault model was originally devised for TTL circuits. It is used for CMOS, although it does not accurately represent all faults, because no better general model has yet been devised.

A stuck-at fault at an input is not necessarily the same as a stuck-at fault at the previous output because there may be a physical break in the connection between them.

For an excellent introduction to the testing of digital circuits and the methods for generating test vectors for both combinational and sequential circuits, see Wilkins (1986).

Fault Simulation and Analysis

The effectiveness of a set of test vectors in detecting the presence of faults is determined by **fault simulation**. This is an extension of the normal computer simulation of a digital circuit but with the circuit description modified to include stuck-at faults on all the circuit nodes. It operates in two stages.

1. For each input vector the simulation is re-run repeatedly with a different fault each time. It compares the actual output of the circuit with the fault-free output each time and, whenever they differ, the fault is listed as being detected by that particular input vector and it does not need to be considered further.
2. The simulations are then repeated for each input vector in turn but only with the faults not covered by previous tests, so that the number of undetected faults gradually falls.

The output of a computer fault simulation gives the fault coverage for the complete set of test vectors. The designer can then go on adding more vectors to the list until satisfied with the total fault coverage.

Design for Testability

An experienced designer can devise a set of test vectors to give a high fault coverage for a small IC without necessarily using formal methods. However, it rapidly gets more difficult to do this as the complexity rises because of the reduced accessibility of deeply embedded nodes. Far greater attention must therefore be given to testability from the very beginning of the design process if the IC is to contain more than a few thousand gates in order to ensure that it can be fully tested in a reasonable time in production.

The problem of testing has been said to be proportional to the number of gates cubed.

We have already seen that partitioning an IC into separate blocks can greatly reduce the number of vectors for exhaustive testing. It also reduces the problem of testing the least accessible nodes by reducing the size of the blocks, and it allows different types of test to be used for memory blocks such as ROM or RAM. Partitioning for test purposes is therefore extremely valuable in designing for testability.

There are many other ways in which digital ICs can be designed to make them easier to test. The most common methods use either **scan paths** or **built-in self-test** circuits.

A scan path can be used for testing any parts of an IC that contain both gates and bistables. All such circuits can be represented as a block of combinational logic, with primary inputs and outputs, and a row of bistables in the feedback paths between the secondary outputs and inputs. A simple example with only two inputs and outputs of each type is shown in Fig. 7.10(a). For testing, the feedback paths are broken at the secondary outputs and the bistables are re-connected to form a shift register when the IC is put into test mode, $T = 1$. This is done quite simply by adding multiplexers to the bistable inputs as shown at (b) in Fig. 7.10.

The shift register scan path has external connections at both ends. In test mode a test sequence is applied serially to the lower end (scan-in) and the output is observed at the top end to check the operation of the bistables. The scan path is then used to apply a known pattern to the secondary inputs, the circuit is put into normal mode, $T = 0$, for one clock cycle, and the secondary outputs of the combinational logic are read out serially from the scan path with $T = 1$. By controlling the primary inputs and observing the primary outputs as well, any

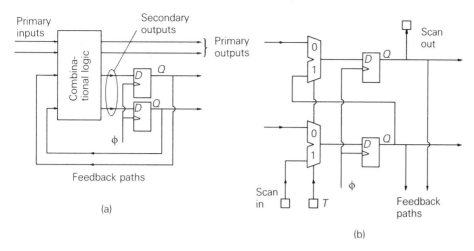

Fig. 7.10 (a) A simple example of a logic block containing both combinational and sequential parts, and (b) the modification of the bistables to include a scan path.

required test pattern can be applied, enabling the combinational logic to be tested as thoroughly as may be required.

In the most common form of built-in self-test (BIST) the bistables used in the normal circuit can be reconnected as either a pseudo-random test-pattern generator, a response accumulator, or a scan path, but with the addition of a considerable amount of extra circuitry. The bistables are grouped into registers called **built-in logic-block observers** (BILBOs) which can automatically carry out complete tests of all the logic when in test mode, and give a binary output **signature** which has the correct value if all the circuits function correctly. Details can be found in Wilkins (1986).

There are many variants of both the scan-path and BIST methods of improving the testability of digital ICs, and it is a subject of growing importance as chip complexity increases. All methods have overheads in the form of extra circuitry which may increase the chip area by up to 20% and decrease the speed. However, this price must be paid wherever chips are to be used in any application requiring high reliability.

Testing for Design Errors

The subjects of the testing and testability of digital ICs are concerned with the detection of faulty chips in production. However, when a new design is fabricated for the first time, the designer is more anxious to find out whether there are errors in the design itself.

It is not always easy to distinguish between fabrication faults and minor design errors, particularly if a small number of prototype chips, say 5 or 10, all behave in different ways. If design errors cannot be detected from these results, there is no alternative but to test a larger number of chips. Although some test vectors may give the correct response even if the IC has a design error, a comprehensive set, designed to give a high coverage for fabrication faults, should also have a high probability of detecting design errors. A design error is distinguished from a fabrication fault (a) by the majority of the test failures occurring for one or more of the same test vectors, and (b) by no chips at all passing all the tests.

Major design errors may be shown up by measuring the supply current taken by the IC and comparing it with expectation. The greatest error is to have a short-circuit across the supply!

123

Design errors are most likely to occur in the connections between cells when they are designed by hand. Common mistakes are to connect a signal line rather than its inverse, or to interchange two connections. These errors should never occur with automatic routing which is controlled directly by the logic netlist. In semi-custom design the most likely source of errors is therefore at the logic design stage although they should have been eliminated there by thorough simulation.

Test vectors are designed to detect stuck-at faults and it is not possible to predict what will happen with a fault-free but logically different circuit design. Unlike fabrication faults it is necessary to locate design errors so that they can be corrected. The type of error in the output, and the vector causing it, may indicate the approximate vicinity of an error but it can take a great deal of effort to locate it precisely and there are no formal ways of going about it. This is one of the reasons why a design interation may cost nearly as much as the original design. The moral is that the designer should do everything possible to ensure a correct first-time design.

Summary

This chapter has described some aspects of digital systems design that are particularly important for ICs. System design starts with a specification and proceeds through behavioural (or functional) and structural levels, preferably using a hardware description language for computer simulation. Simulation is essential for verifying the correct operation at each stage and particularly at the gate level where realistic timing is introduced. Synchronous operation using single- or two-phase clocks is almost essential for complex designs. Circuit delays determine the maximum clock frequency for the IC and design must minimize their effect. Some circuit precautions are suggested for reliable design in high-speed operation.

The testing of ICs in production can raise many problems. The strategy for testing must be considered throughout the design process. Models of faults can be included in simulations used to determine the effectiveness of sequences of test vectors. Testability can be improved by partitioning the chip for test purposes and by including scan paths and self-testing structures.

Some semiconductor companies use electron beams as probes to find the voltage at nodes deep inside an IC, which is a great help in locating design errors. Mechanical probes can only be used if extra pads have been provided on the chip as they are normally too coarse to make contact to ordinary circuit nodes without damaging them.

Practical Cell-based IC Design 8

After studying this chapter you should be able

☐ To recognize what is meant by 'good' design practice.
☐ To understand the general features of the CAD hardware and software required. An integrated set of CAD tools providing a route to silicon fabrication helps to achieve good design practice.
☐ To understand the role of cells and cell libraries in the design of semi-custom digital ICs.
☐ To know what fundamental decisions have to be taken by the design manager before a design project can start.
☐ To start doing practical design of semi-custom ICs down to the cell level. To understand the role of simulation, and the features of some commonly used CAD packages.
☐ To appreciate the simplicity of cell placement and autorouting for the completion of semi-custom ICs.

The previous chapter established the general background for digital IC design by considering simulation, timing and testing. In this chapter we are concerned with practical methods for designing ICs down to the cell level with emphasis on semi-custom design, and in Chapter 9 we will describe the additional stages, involving circuits and layout, for the internal design of cells and for full-custom ICs.

Design Practice

'Good' Design

Design is an open-ended creative process that needs a keen judgement to balance out conflicting demands such as those between performance and product cost or design time. For this reason, design problems never have a uniquely 'best' solution although we can usually identify what would be expected of a 'good' design. Due to the large number of trade-offs made in a large design there is nearly always scope for improving it after the first attempt.

We have already considered some of the features of 'good' IC design practice in Chapter 1. They may be re-stated as follows.

1. Low design time and cost. Particularly important for ASICs to be used in competitive products.
2. Robust design. To ensure that the IC will work correctly in spite of wide manufacturing tolerances. In particular, its operation should not depend on particular delay values.
3. Verified design. To give a high level of confidence that the IC will work correctly by simulating it as thoroughly as possible.
4. High-performance design. To achieve high-speed operation and the minimum

Low design time is essential and not just to reduce the cost. It has been estimated that about half the ASICs designed in industry are never manufactured because of changing commercial priorities before the product is fully developed.

possible chip size by using the silicon and circuit technologies to the best advantage so as to reduce the production cost.

5. Testable design. To ensure that the chip can be thoroughly and rapidly tested in production.

The complexity of digital ICs raises some very fundamental problems about design even for chips containing only a few thousand gates. The problems are greatly reduced, and good design practice is encouraged, by making full use of the appropriate CAD tools.

The Uses of CAD Tools

It is nowadays absolutely essential to use a computer for any IC design. In addition to a range of CAD tools, the computer can provide a complete environment for design in the form of a database. This contains all the files defining the various parts of the design as they are created and evolved, and it enables them to be linked together when appropriate. To be most effective, the individual tools should be integrated into the complete CAD system in this way.

CAD is used for:

1. Functional and logic simulation down to the cell level of design as described in Chapter 7;
2. Circuit simulation, which we will come to in the next chapter;
3. Testability analysis and automatic test pattern generation;
4. Layout, which consists of the placement and routing of cells in semi-custom design and mask layout in full-custom, both of which will be described later; and
5. Post-layout verification that will also be described later in this chapter and in the next.

CAD Hardware

In principle almost any computer system can be used for CAD although the availability of software may in practice limit the choice. Most CAD software packages support graphical displays of circuit diagrams, waveforms, and/or layout. Although such graphics look very nice they are by no means essential for semi-custom design since all the design information, including layout, can be contained in text files, which saves the expense of graphics terminals. A graphical display, preferably in colour, is however necessary for full-custom layout. If graphics are to be used, a compatible plotter is essential to get paper output for inclusion with the written documentation that should accompany any design project.

All the CAD tasks for IC design require a considerable amount of computer power if they are not to be impossibly tedious to use. The simulation – circuit-change – re-simulation cycle in particular must be as rapid as possible so as not to waste the designer's time which is a most valuable resource. For small designs, typically masked ASICs or PLDs with less than 2000 gates, a fairly powerful PC can be adequate for design. Any larger design, however, must be done on a workstation, such as an Apollo, Sun or Vaxstation, or a multi-user mini-computer. The advantage of PCs and workstations is that the graphics are an integral part of the system whereas a minicomputer requires separate graphics drivers which may then restrict the choice of terminals. For large designs containing upwards of 10 000 gates it can take many hours for any of these machines to carry out a full simulation at gate level or to autoroute a gate array. Design offices doing this scale of design usually use a mainframe computer for such tasks.

Whatever type of CAD hardware is used it must obviously provide a suitable output for extracting the final design data and sending it to a semiconductor manufacturer. The physical transfer of a tape or magnetic cartridge is usually safer than electronic file transfer for the communication of such a large amount of information. The details of the data format must, of course, be agreed between both parties beforehand.

CAD Software

CAD software varies between simple easy-to-use systems for small semi-custom designs and highly sophisticated packages for complex ICs. A complete software system always contains tools for simulation and for layout and it is usually supported by a range of database utilities, and schematic capture and other display programs. Semi-custom CAD packages make use of data contained in the cell library as described on page 129. All the data files must therefore be in a compatible format and easily accessed so that the software modules can be fully integrated into a complete and easily used design system.

Nearly all semiconductor companies produce gate-array and standard-cell ASICs, and they invariably have their own semi-custom CAD systems. Manufacturers' software always contains all the necessary library and chip information, and it provides a very reliable route to silicon fabrication, although necessarily limited to a single manufacturer.

More general-purpose software is produced by CAD companies outside the semiconductor industry and it can also be used to design semi-custom ICs. However, in this case it is necessary for the designer to obtain either a general-purpose or a manufacturer's cell library, to link it in to the simulator and layout tools, and to ensure that the final output is acceptable for fabrication. General-purpose tools often provide more powerful features than manufacturer's software but at the price of awkward interfaces to other tools which may make them more difficult to use. There are many potential hazards in trying to link tools and libraries from different sources and the designer is well advised only to use a CAD system that has previously been proved to work right through to chip fabrication. The best in this respect have the hardware and software integrated in a workstation as in the Daisy, Mentor and Valid CAD systems. The extra expense of such computer systems can be justified if they are to be entirely dedicated to electronics CAD.

In addition to the technical capabilities of CAD software, it is important that it should be user friendly so that the designer can concentrate on design rather than computing. Unfortunately, some familiarity with computer notation is highly desirable, however. It is inevitable that many files and directories will be used in an IC design of any complexity, and the naming conventions must be understood so that they become self-explanatory. The only interaction with the computer operating system should then be for editing files. Program control should be built into the applications software itself, preferably through easily understood menus. The clarity of the software documentation and on-line help messages can also make a great deal of difference to the designer's productivity.

Some CAD companies produce integrated hardware—software systems which overcome the problems of installing software on workstations, but at increased cost.

This problem should be reduced in the future by the development of standard interfaces using the electronic design interchange format (EDIF).

Full-custom and Semi-custom Design

We have seen that top-down digital systems design starts from a basic idea, 'We want a chip to do....' The requirement is expressed more precisely by writing a formal specification and devising a block diagram of the IC to give an overall idea about how it will work. The design then proceeds by partitioning each of the main blocks into smaller and smaller sub-blocks until they can be identified with register-level digital hardware. The procedure below this level depends on the important choice between the alternatives of full-custom and semi-custom design which were described briefly in Chapter 1.

We will define these two main approaches to the lower levels of design more formally as follows.

1. *Full custom.* Design going through all the remaining levels including cell design, circuits, and layout.
2. *Semi-custom.* Design going down to the cell level only, avoiding the need for circuit design by using a library of pre-defined parts.

Full-custom gives the better performance because the transistor-level circuit design can be optimized for the highest operating speed and the layout can be made very compact. However it requires a considerably greater time than semi-custom design and is more likely to contain errors, both of which substantially increase the cost.

Semi-custom design can be done rapidly and with a high probability of success if simulation is used at each stage to verify that it is correct before proceeding to the next level down. The elimination of the lower levels of design in semi-custom design together with the development of powerful but easily used CAD tools, have therefore been spectacularly successful in overcoming the problems of ASIC design and opening up silicon chip technology to the entire electronics industry.

To appreciate the reasons for the different benefits of full- and semi-custom design we need to consider the use and properties of cells in both styles of design.

The Role of Cells in IC Design

A cell is a small piece of circuitry used as a component in the layout of an IC. It is a physical rather than an abstract element with a shape, usually rectangular, and a boundary of certain dimensions. The layout of the register-level blocks, and hence of the entire chip, is built up from cells. For digital ICs they typically contain gates and bistables in commonly used groupings. A cell can contain anything between a single inverter and a logic circuit equivalent to about 10 or more gates. We will describe some analogue cells in Chapter 10.

In semi-custom design, cells are provided in the cell library, but in full-custom design they are made up by the designer, which gives far greater flexibility. We will consider cell design and full-custom design in Chapter 9.

In doing top-down design the basic circuit function is broken down in stages to the register level. This is where attention is first given to the type of cells that might be used on the silicon itself. In semi-custom design, the cell library is consulted to look for suitable cells to build up into the register level blocks. In full-custom the designer tries to identify a small number of cell types that can be used repeatedly to build the blocks in much the same way. In both cases the logic design of the register-level blocks is then directed towards using these particular cell types. This is really a bottom-up process because it synthesizes large blocks from smaller ones whereas the top-down stages are concerned with ways of splitting up the larger blocks into sub-blocks.

The logic level in IC design should always use a clearly defined selection of cell types. When a design is completed at this level, the physical layout can be built up by assembling the cells and their interconnections as we will see later. The cells therefore provide the link between circuit design and layout.

Cell Libraries

Cell libraries are provided by most semiconductor companies for use with their semi-custom digital ASIC products such as gate arrays and standard cell chips. In addition, some CAD companies produce general-purpose libraries for ICs that might be fabricated by any semiconductor company having the right technology.

All digital functions can be built up from about 40 basic cell types, but some libraries also provide higher-level groupings, or **macros**, leading to several hundred types. The basic cells always include:

1. Inverters with various drive capabilities;
2. NAND and NOR gates with various numbers of inputs;
3. Complex gates such as AND–OR–INVERT;
4. Multiplexers;
5. Bistables of all types with and without reset and clear inputs; and
6. Pad circuits for inputs and outputs with interfaces to external logic.

The pad circuit at the input contains diodes and a resistor to protect the chip from excessive input voltages. The output pad circuit is essentially a power amplifier for driving the external load from a very small on-chip circuit.

The higher-level macros are of many types, but they may include a wide range of register, counter and arithmetic elements, decoders, and multi-level logic cells. Much of the labour of logic design can be eliminated by using these high-level cells.

The cell library information is provided in three parts.

1. Information for the designer, contained in a printed design manual;
2. Information for the CAD system, usually containing graphical symbols for schematic capture, simulation models and cell-outline geometry; and
3. Information for fabrication which requires the complete layout of each cell. This is usually added to a design by the semiconductor manufacturer and it is not usually available to the designer.

The information provided for the designer always includes:

1. A precise description of the cell parameters in the form of a logic circuit and truth table;
2. The electrical parameters of the cell; and
3. The cell size, not necessarily in microns, but in terms of a certain number of layout elements.

Some libraries also include transistor circuits for cells and the layout wiring for gate arrays even though they are not needed for design using modern CAD tools.

The electrical parameters of the cells include their driving capability, input loading, delay values, and power consumption. The driving capability of a cell is expressed in terms of a number of input **load units**. In MOS, a load unit is the input capacitance of a minimum size inverter, and values are given for the inputs of all the cells. The drive capability of gates is often four or eight load units so that buffer cells have to be used to drive higher loads. The buffers themselves usually have an input load of more than one unit.

The delay between the input and output of a logic circuit depends on the electrical load as seen in Chapters 4 and 5. The load is determined mostly by the fan-out. Delay values are therefore given in the library information for a circuit with a fan-out of one and for the additional delay for every extra load unit. It is also quite common to give maximum and minimum as well as typical values for the delays for both the rising and falling signal edges.

In MOS circuits, the capacitance of the interconnections may also add appreciably to the total load and hence to the delay. This capacitance is not known until the length of the interconnections can be determined from the layout. Semi-custom CAD layout packages usually contain a program to calculate the capacitive load of each track in the final layout and add it to the fan-out load of each cell. The circuit is then re-simulated to ensure that the increased delay values do not lead to incorrect operation. For this to be done the CAD system must be provided with the interconnect capacitance per unit length and data on how the cell delays are affected by the extra loading, and this information may or may not be provided to the designer also.

The power consumption of each cell must be known to ensure that the maximum recommended dissipation of the chip is not exceeded. For MOS circuits, data is given for calculating the power as a function of frequency, and for bipolar the additional static power dissipation is also given.

All this data on the electrical properties of the cells is usually given for room temperature and a stated supply voltage, usually 5 V. Scaling factors are given for the effects on all the electrical parameters of higher and lower temperatures and supply voltages.

Fundamental Design Decisions

Now that we have described the general features of the CAD tools and cell libraries that contribute greatly to good design practice, we can start to consider a complete design project.

Some fundamental decisions need to be made before starting the detailed design of any product. The success or failure of a company manufacturing electronic equipment can easily depend on making the right decisions on the technology to be used, so that they must be taken by a senior design manager. These decisions also determine the design methods so that they have a great impact on the designer!

An IC is only one component in a digital system which might be built using standard ICs, microprocessors, ASICs, or any mixture of these types of chip. We will assume here that ASICs are considered to provide the most economical way of producing a sub-system of the required performance, so we must consider the next level of decisions that have to be taken before contemplating the design. These are:

1. The choice between one or more ASICs;
2. The choice between a fabricated (or 'masked') ASIC and a programmable device (PLD);

The drive capability of bistables is often only 1 or 2 and buffers often need to be added to the Q and $\bar{Q}$ outputs.

This simulation after load 'extraction' is called **post-layout simulation** or **back annotation**.

For CMOS the delays increase by about 0.3% per kelvin rise in temperature, and by about 25%/V fall in the supply voltage.

Decisions 1–2 are not necessarily made in this order as they are all linked. They also depend very much on the size of design being contemplated. Decisions 4–6 are more sequential but they all require the collection of commercial information on cost and performance from semiconductor companies.

3. The choice between full-custom, and semi-custom gate array or standard-cell IC if it is to be a fabricated chip;
4. The choice of circuit technology, MOS or bipolar, and the process details;
5. The selection of the silicon manufacturer and fabrication process; and
6. The choice of CAD tools.

These decisions must be based on a detailed examination of the required chip performance and analysis of the design and production costs for all the alternatives. Often this will have been done for some previous product leading to investment in a particular CAD system which may limit the choices for later products. The IC requirements can be summarized by the number of gate equivalents on the chip, the total number of input, output, and supply connections, and the operating frequency. If the ASIC is to replace an existing printed circuit board of standard ICs these requirements are known accurately at the outset, but for an entirely new design the figures have to be estimated with sufficient accuracy to establish at least the relative costs of the alternatives. Excessive figures for any of the requirements, such as a very high speed or complexity, can easily narrow down the choices listed above.

For all the alternatives, the cost will also depend greatly on the number of chips likely to be produced. The costs of the investment in CAD tools and of the design work have to be included, giving the overall trends already shown in Fig. 6.12. A final factor that has to be taken into account in determining the type of chip to be used is the likelihood of specification changes in the later development of the complete system. Changes need expensive design iterations for fixed-function ASICs but they can be made almost instantly in programmable ICs.

For any IC design to be fabricated, the interface to the silicon manufacturer, or **vendor**, also has to be formalized before starting the design. We have seen that all styles of design prevent the designer from working at the lower levels, but a definite cut-off of this type is only possible if there is a clear division of responsibility between the designer and the semiconductor manufacturer, and this requires a well-defined, almost legal, interface between them. The designer must be provided with clear and precise information about the chip performance at the lower levels. For semi-custom design this is given in the manufacturer's cell library and data on chip architectures. For full-custom design the manufacturer must provide the electrical properties of the transistors and layers formed on the silicon, together with the layout design rules for the fabrication process to be used, so that workable circuits can be designed.

Other matters that have to be agreed between the designer and manufacturer of any fabricated IC are the design verification procedures, the format for the physical transfer of the final design data to fabrication, the type of packaging to be used, and arrangements for testing the final chips. The manufacturer usually checks the processing by making measurements on 'drop-ins' (Fig. 2.1) and makes the functional tests agreed with the designer. These agreements are to the benefit of both parties and they should lead to the smooth production of working ASIC chips.

All these decisions may require a considerable effort before the design of an ASIC can start in earnest, but they are essential for a satisfactory and economic conclusion. They can well influence the choice of CAD tools which may have to be supplied by the semiconductor manufacturer with whom the designer works closely. Although a wide-ranging investigation of alternatives should always be

A company manufacturing ASICs is sometimes called a **silicon foundry** which gives the impression that the chips are stamped out! It is sometimes possible to get silicon designs fabricated through a **silicon broker** who selects the best foundry for a given design and procures the chips on behalf of the designer.

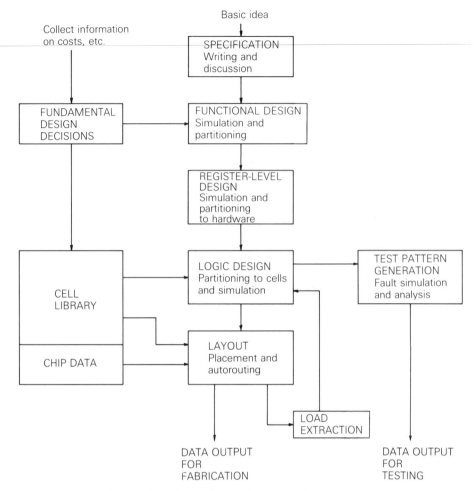

Fig. 8.1 The main semi-custom design stages.

made, the ultimate choice of ASICs for many products at the present time will be one of the many types of gate-array or standard-cell chips, usually in CMOS, that are manufactured by almost every semiconductor company in the world, and we will follow this particular path for the remainder of this chapter.

Semi-custom Design Flow

The seven main stages in semi-custom design are shown in Fig. 8.1 which also indicates where the fundamental design decisions and library data are fed in. For small ICs the specification and functional operation may already be given so that these stages can be omitted as in the Design Example at the end of this chapter.

At the end of each stage the data on the design is contained in a new set of drawings, lists, and simulation results mostly held in computer files in the CAD database but supplemented by the designer's written notes and sketches. The type of information generated at each stage is shown in Fig. 8.2. The design data divides into many different parts as the IC is developed down to the cell level, but it finally comes together again in the chip layout.

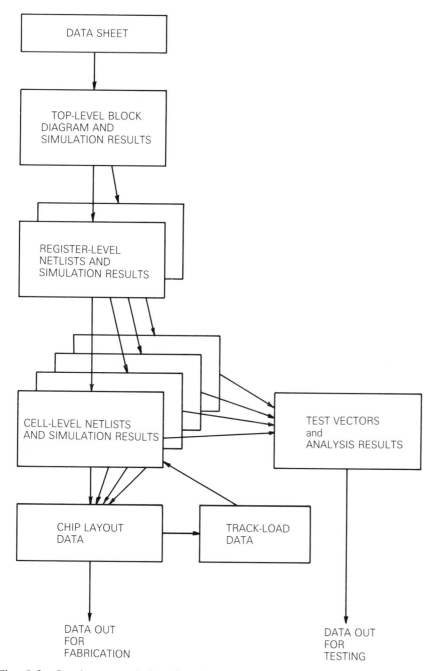

Fig. 8.2   Semi-custom design flow showing the data generated at each stage.

We will briefly describe what has to be done in each of the seven main design stages.

Specification

The specification of an IC is a detailed written description of its overall function, giving the relationship between all the input and output data signals, and the

The pad circuit at the input contains diodes and a resistor to protect the chip from excessive input voltages. The output pad circuit is essentially a power amplifier for driving the external load from a very small on-chip circuit.

Imprecise definition of chip function is one of the main sources of error in IC design.

effects of the various control bits. It usually contains diagrams showing the relative timing of data bits, the clock and the control bits. The specifications of standard ICs are given in semiconductor data books and it is a good idea to have a look at some of these before starting a new design. Fully detailed definitions of the digital signals and their functions are far more important to the designer than physical characteristics, such as the effect of temperature or power supply variations, which are in any case determined by the fabrication technology.

The detailed specification should be discussed with systems designers who might be interested in using the chip. The interfaces determining the transfer of data and control signals into and out of the chip from a synchronous or asynchronous environment have to be agreed. Details such as the polarity (logic 0 or 1) of resets and other controls and the clock edges on which they become active, are extremely important.

Functional Design

The functional design develops naturally from the specification by refining ideas on how the IC will work. These are generally expressed in the form of block diagrams containing functions such as logic, memory, and control units. A formal design method should generally be used and this starts by considering the stages in the processing of data by the IC as expressed in a flow chart or algorithm written in an HDL. For very simple designs, however, the block diagram may immediately be apparent or it may be deduced intuitively as described in Chapter 7. In other cases it may be given by the requirement, as when an existing board of standard ICs is to be replaced by an ASIC, for example.

The signal flow between the blocks must be defined precisely in much the same way as in the specification of the complete chip. Unless the IC is very simple it is always worth simulating its functional behaviour using a high-level simulation package. If it does not work correctly, adjustments can be made to the functions of the blocks or to the interconnections, it being far easier to make changes at this level than later because of the comparative simplicity of the functional descriptions. High-level simulation is also useful for comparing different ways of implementing the specification and it can lead to great improvements in the performance of very complex ICs such as microprocessors.

The functional design of a complicated chip continues by partitioning the block diagram into smaller and smaller units and adding further detail about its digital operation. In doing this the designer has in mind the types of digital functions that can normally be made, examples being given in Appendix 5. Computer simulation is used at every step to make sure that each division of the blocks gives the same overall behaviour. The design of a large IC should only proceed if the correct operation has been completely verified by thorough functional simulation in this way.

For small designs the functional level of design can usually be omitted because a register-level hardware description is sufficiently obvious from the specification.

Register-level Design

The conversion of a functional diagram into a hardware block diagram at the register level has been described in Chapter 7. The register-level blocks include all the functions familiar to digital designers such as data registers, many types of counters, adders, decoders, multiplexers, etc., that will become identifiable circuit blocks on the final chip. For such blocks the conversion involves changing a functional description to a hardware description in a simulation model file.

Some functional blocks, such as combinational and control functions, are converted by using classical digital design techniques based on Boolean algebra and Karnaugh maps for logic minimization, and state machine methods for sequential design. These methods give circuits made up of gates and bistables that can be checked by simulation.

The conversion of the design into digital hardware is straightforward if the partitioning has been directed towards an eventual hardware implementation. Some simulators allow blocks to be specified by either functional or hardware models which is most useful for checking that the behaviour has not been changed by the conversion process.

Logic Design

In semi-custom design the register-level blocks are built up from library cells. The amount of work involved in the logic level of design therefore depends greatly on the contents of the cell library that is used. As we have seen, some of the larger libraries contain functions such as registers or counters which may be used directly as the main components in the design. More commonly, however, the larger hardware blocks have to be built up from cells containing gates and bistables, which is slightly more time consuming.

Simulation is used again to confirm that the main circuit blocks work correctly when they are replaced by interconnected cells, and, if they do, to verify the operation of the complete IC. Cell-level simulation usually includes the delays which are adjusted automatically for fan-out in many CAD systems.

The cell-level simulation should initially use a fairly low clock frequency. If the circuit works correctly, the clock frequency can then be increased until errors are introduced due to delays, and this enables the maximum frequency of operation to be determined. If this is done for each of the major blocks before simulating the complete IC, the slowest path can be identified and attempts made to re-design it so as to increase the speed of the entire chip. Delays may be reduced by giving careful consideration to the slower circuits and eliminating excessive fan-out or multi-level logic for example.

Test Vector Generation

The problems of testing ICs were discussed in Chapter 7. The designer is responsible for producing an efficient set of test vectors and some semi-custom chip manufacturers will only accept a design for fabrication if the test sequence gives a high fault coverage (such as 95%). Many semi-custom CAD packages include fault analysis for evaluating test patterns. A few simulation packages also include automatic test pattern generation (ATPG), which gives 100% fault coverage for combinational logic, while still leaving the problem of test pattern generation for the sequential logic to the designer.

There is no reason why test features such as scan paths cannot be included in semi-custom ICs although they take up extra area. Some of the more advanced cell libraries include bistables with built-in scan paths which can greatly improve the testability.

Placement and Routing

The layout stage of semi-custom design consists of deciding where each cell will be placed on the chip, and then determining the paths of the interconnecting tracks. Both jobs are done automatically in many CAD systems although with some interaction by the designer.

The cell placement tries to produce a compact layout fitting closely into the plan view of the chosen chip. The chip size is determined either by the number of input and output pads in the design or by the number and size of its cells. In most semi-custom layout the actual utilization of the silicon area is poor due to the routing channels and it is made worse with gate arrays by the inability to use all the cells on a chip, 80% utilization being quite a high figure.

In deciding the cell positions, closely related parts are grouped together to reduce the lengths of the interconnecting tracks. Every track has capacitance which adds to the total load driven by the cell outputs, so increasing the circuit delays. Short connections in critical parts of the circuit can therefore improve the speed of operation.

The top level of chip layout is called 'floorplanning'. It is very important in full-custom design (Chapter 9) but it also needs to be considered in the positioning of groups of cells in semi-custom layout.

The placement/routing software needs to contain details of the chip architecture and the layout design rules for the interconnections. The cell positions are determined by the placement program perhaps with manual intervention by the designer to place some of the critical cells. This is done either by drawing on a graphics screen or by typing in a list of coordinates.

An **autorouting** program is then used to determine the positions of the interconnections, as given in the cell netlist. If the program is unable to place all the tracks the designer may have to complete the job manually, which is not particularly tedious. The final CAD output is in the form of a file giving the coordinates of all the cells and tracks expressed in a format that can be used to create the masks for fabrication.

Routing channels can get congested in gate arrays because their width is fixed and the autorouter may not be able to fit in all the tracks. In a standard-cell IC, however, the gaps between the rows of cells can be varied to accommodate any number of tracks and 100% autorouting should be obtained.

Autorouting saves a great deal of design time but, more particularly, it ensures that the physical interconnections must be identical to the circuit netlist at the logic level, so eliminating some of the most common sources of error in chip design.

Post-layout Simulation

A final check of the correctness of the design is made by re-simulating it after layout. Many CAD systems can evaluate the track capacitances automatically and add them to the cell outputs so that more realistic values of the delays can be used in the final simulation. If the design fails this test it is necessary to change the cell placement to reduce the lengths of critical interconnections, or, failing this, to go back to an even earlier stage and adjust the design at the logic level.

CAD Tools for Cell-based Design

It is not the intention of this book to describe the use of particular CAD tools, but rather to present the background of IC design applicable to all of them. The tools themselves change rapidly as manufacturers regularly produce new versions of the software, and most of them are adequately described in the manuals. However, it is worth describing the main features of a few of the tools commonly used in universities and polytechnics in the UK. Most of this software is provided centrally through the UK Higher Education Electronics Computer Aided Design (ECAD) Initiative.

Details and practical examples of the use of ECAD tools are given in Jones and Buckley (1989).

For semi-custom design, company-specific software provides the most certain route from basic idea to silicon fabrication because the semiconductor company providing it wants to ensure customer satisfaction. Such software covers all the features required for reliable design using the company's particular technology, such as the cell library, chip outline, and all the necessary verification tools for either gate arrays or standard-cell ASICs. Examples are:

1. BX   produced by Micro Circuit Engineering Ltd,
2. Quickchip   produced by Qudos Ltd for Texas Instruments gate arrays,
3. Solo   produced by European Silicon Structures Ltd, and
4. PDS   produced by Plessey Semiconductors Ltd.

In addition, the design systems for programmable ICs, such as PLDs and programmable arrays, are very similar but their output is used to program the chip directly.

General-purpose software can be used for any style of design or technology and it generally includes more powerful features. Three of the commonly available general-purpose systems have hardware description languages for writing behavioural and structural descriptions that can be expanded down to the cell level:

1. ELLA   produced by Praxis Ltd,
2. HELIX   produced by Silvar-Lisco Ltd,
3. HILO   produced by GenRad, Inc.

To use any of these for semi-custom design it must be linked to an appropriate cell library and layout software obtained from a different source.

Among the company-specific systems, Solo, mentioned above, also includes a behavioural HDL in its MODEL language.

ELLA

ELLA is a design system that can be used anywhere between the specification and gate levels. It is based on the ELLA language for describing signals on 'wires' between the nodes of functional or hardware blocks. The language is simple and flexible and it can be used throughout the hierarchy of levels in top-down design from behavioural modelling, for evaluating system architectures, down to register-level and gate hardware. Simulation can be done at any level or with mixed descriptions.

HELIX

HELIX is a behavioural modelling language that can be used to cover the same levels of design as ELLA. The models of the system blocks are written in HHDL (hierarchical hardware description language) which is like an extended version of Pascal so that it is very flexible. The system architecture is usually described graphically using the CASS schematic entry package, produced by the same company, with either user-defined or library symbols linked to the models. There is no need for a separate HELIX simulator because the Pascal descriptions are already procedures so that the system is simulated simply by running the program.

HILO

HILO is fundamentally a powerful logic simulator. System elements can be described either **functionally**, enabling register-level blocks to be modelled concisely, or **structurally** in terms of gate elements. The connections are easily described as text or by conversion from a CASS schematic. HILO can allow for variable tolerances in circuit delays and for the different signal strengths of circuit outputs. It can include automatic test pattern generation and sophisticated fault-analysis features.

More Advanced Cell-based CAD Tools

The semi-custom design methods described above can be enhanced by using more advanced CAD tools particularly for large designs. Reference has been made in Chapter 6 to CAD systems for combining cells into regular blocks such as ROM and RAM for inclusion in standard-cell chips or channel-less arrays. Any circuit block composed of repeating cells is well suited to automatic generation from a behavioural or structural description written in an appropriate HDL. Some of the more advanced CAD systems enable circuit blocks to be **compiled** in this way.

To compose a block automatically the system must know the size of block required, for example the number of inputs and outputs, and whether any options are to be included. These **parameters** are specified by the designer and the system produces the outline geometry of cells assembled into a block, a symbol for graphical schematic entry, and a simulation model for the block. These can be used with data on other blocks and with library cells to build up the complete design as in any semi-custom method.

An example of this type of automatic design using such **parameterized** cells would be the generation of 4-, 8-, 16-, 24-, or 32-bit adders with carry-lookahead every four bits, and with the options of subtraction and/or a registered output. Other examples are blocks such as PLAs, arithmetic units, register stacks, counters and multi-way multiplexers any of which may be compiled from a few textual commands. The software for doing this is often produced by semiconductor companies for their own particular technologies.

Automatic block generation is the most effective way of designing large ICs economically. When combined with advanced standard-cell or channel-less array technologies it can be used to produce very dense circuits with extremely high performance.

Summary

This chapter has described the stages involved in cell-based semi-custom design and the characteristics of CAD hardware and software. The library information contains full details of the performance, size, simulation model, and symbol for each cell. After choosing suitable chip and circuit technologies for an ASIC, the design process starts from the specification and ends with a chip layout complete with the interconnections between cells.

This design process is a mixture of top-down and bottom-up design meeting at the register level. Pure top-down design is an impossibility because it always requires

PLAs and finite-state machines, specified by Boolean equations and state tables, respectively, are two of the most useful blocks that can be generated automatically with quite straightforward software.

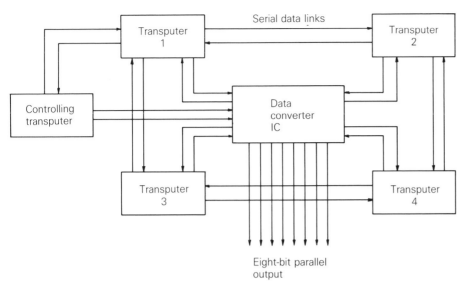

Serial data links

Transputer 1

Transputer 2

Controlling transputer

Data converter IC

Transputer 3

Transputer 4

Eight-bit parallel output

Fig. 8.3   Typical application of the data converter IC in the Design Example.

knowledge of where it is going and, in any case, practical design is iterative, continuously going up and down between levels in an attempt to find a best solution.

Design Problem: A Transputer Data Converter

The transputer is an advanced IC designed for parallel processing applications. For communication between the transputers in a system, each of them has four serial links consisting of an output for transmitting data and an input for receiving data and acknowledge signals. The exercise is to design an IC to convert the output from one of four transputers to a parallel form for an external connection as shown in Fig. 8.3.

Specification

Serial data is transmitted on the links between transputers in eight-bit bytes. Each byte is preceded by two start bits and followed by one stop bit. When a byte has been received, an acknowledge pulse (ACK) must be sent back to the transmitter with the timing shown in Fig. 8.4.

The IC is to select one of four inputs and outputs, LIN1–4 and LOUT1–4, as determined by two select bits, SL1 and SL2, set by the controlling transputer:

1. LIN1 and LOUT1 selected by SL1 = 0, SL2 = 0,
2. LIN2 and LOUT2 selected by SL1 = 0, SL2 = 1,
3. LIN3 and LOUT3 selected by SL1 = 1, SL2 = 0,
4. LIN4 and LOUT4 selected by SL1 = 1, SL2 = 1.

The start bits must be removed so that only the data is transferred to the parallel output register. When the output data is valid a data-ready output, DRDY, is set high, and the peripheral equipment is triggered on the rising edge of DRDY to read to parallel output. If a stop bit is not received, the ACK pulse is not sent, DRDY remains low, and the converter will wait for a new serial word. The power-on reset input, RST, active-low, sets the converter to the wait state. The

Fig. 8.4   Timing diagram for the data converter IC.

input data is at a bit rate of 20 MHz synchronized with the system clock. The input/output pin names are shown in Fig. 8.5.

Design method

A block diagram for the IC (Fig. 8.6) is so obvious that there is no need for functional simulation. The IC has three internal control signals:

1. SIN – serial input to shift register and the same as the data from the selected link;
2. SEN – shift register enable, active high. If SEN = 1 new data is shifted in. If SEN = 0 no shift.

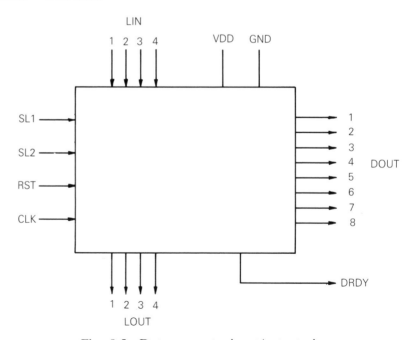

Fig. 8.5   Data converter input/output pins.

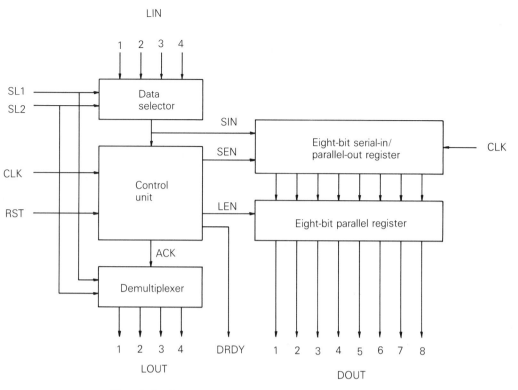

LIN

1 2 3 4

SL1 —— Data selector

SL2 ——

SIN

SEN

CLK ——

RST ——

Control unit

Eight-bit serial-in/ parallel-out register

—— CLK

LEN

Eight-bit parallel register

ACK

Demultiplexer

1 2 3 4   DRDY   1 2 3 4 5 6 7 8

LOUT

DOUT

Fig. 8.6 Data converter block diagram.

3. LEN – latch enable, active high.

The central part of the IC is the controller which is a finite-state machine with one input, SIN, and four outputs, SEN, LEN, DRDY, ACK, plus a reset. The machine has 13 states and its sequence of operation should be:

1. Wait for the first start bit.
2. When the first start bit is received, look for the second start bit in the next clock cycle. If present, set SEN = 1. If not present, return to wait state.
3. Keep SEN = 1 for eight clock cycles.
4. Set DRDY = 0, SEN = 0, and look for stop bit = 0 on next clock cycle. If present, set LEN = 1 and go to 5. If not present, return to wait state.
5. Set ACK = 1 and DRDY = 0 on next clock cycle.
6. Set ACK = 0, DRDY = 1, and return to wait state on next clock cycle.

As an exercise, this specification can be written more formally, as a Pascal procedure, for example.

To design the controller:

1. Convert this description to a state diagram,
2. Draw a state table and assign codes to each state,
3. Derive equations for the inputs of the four D-type bistables that will be used and the corresponding outputs.

Design the controller and the other blocks using cells from whatever CAD system is being used. The D-types in the controller should be positive-edge

141

triggered, but transparent D-types can be used for the registers. The data selector and demultiplexer are identical.

Simulate the individual parts and then the complete circuit. Find the maximum clock frequency. Even if it is greater than the 20 MHz specified, try to increase it by redesign for the Mark II chip.

Devise a set of test vectors to give 95% fault coverage.

Lay out and route the cells for the semi-custom technology being used. Alternatively, it could be implemented as a full-custom design.

Postscript

This is an extremely simple IC but it is very suitable for a first attempt at design. The quality of the design will be shown by the speed at which it can be made to work, the number of cells used, and the compactness of the layout.

Full-custom Circuit and Layout Design

9

Studying this chapter should enable you

Objectives

☐ To appreciate the limited role of full-custom IC design in industry.
☐ To understand the stages in full-custom design and why it is so much more expensive than semi-custom design.
☐ To see how the extra flexibility in logic and circuit design is used in full-custom design and how these stages interact.
☐ To understand the value of designing the floorplan of an IC to give a regular flow of data and control signals.
☐ To see the need for circuit simulation in IC design and the features of a simulation package such as SPICE. The importance of accurate SPICE parameters must be appreciated.
☐ To understand how circuit cells are converted to layout and the need for design rules. The cells and interconnections must fit together to cover the entire active area of the floorplan.
☐ To appreciate the problems of layout verification.

We use the term 'full-custom' for IC design that goes down to the levels concerned with transistor circuits and their layout without using predefined cells. The advantages of full-custom over semi-custom design are that it gives:

1. A smaller chip area which reduces the chip cost in large-scale production,
2. The possibility of a considerably higher operating speed, and
3. Freedom for the designer to use novel circuits for critical applications and array structures such as PLAs.

The disadvantages of full-custom design are the far greater design time and cost. This is made even worse by the strong possibility that the ICs will not work correctly at the first attempt, usually because of layout errors. It then requires one or more iterations to correct the design, with additional mask costs and delays. Verification of the accuracy of a layout is a major task for full-custom CAD software. Although it greatly increases the chance of success the cost of full-custom design remains extremely high.

Because of these major disadvantages, full-custom design is not used for ASICs and it is rarely done even in the semiconductor industry. It is also unnecessary because nearly all ICs are developments of earlier ones or they use parts of previous designs. A well-organized industrial design office keeps a database containing the results of all previous design projects. Parts of these can be modified to have slightly different functions or for a different production technology and this takes far less time than starting a design from scratch. The modified parts are assembled, perhaps with some genuinely new full-custom blocks, to create a new design. This style of design is done mostly in the semiconductor industry itself because the cost, although reduced compared with a completely new design, is justified only by large-scale production.

The role of full-custom design in industry is therefore in:

1. The modification of previous circuits,
2. The creation of new blocks to interface with existing circuits, and
3. The design of cells for semi-custom libraries.

Most of these activities are concerned with the circuits and layout of cells rather than complete ICs so that it is important that all designers should have some knowledge of cell design.

Practical full-custom design also has a very important role in electronics education because it provides many illustrations of the strong interactions between all levels of design that are hidden in semi-custom methods. Courses for professional designers therefore often include large, full-chip, full-custom design exercises of a type that is seldom done in industry. We will consider the design flow for such a 'pure' form of design even though it is rather artificial.

Full-custom Design Flow

Figure 9.1 shows the main stages in full-custom design. It should be compared with the semi-custom design flow in Fig. 8.1.

The same fundamental design decisions have to be taken in the two styles of design, and the top levels, concerned with the specification and functional and register level design, are identical and they use the same types of CAD tool.

Working down the main sequence in Fig. 9.1, changes first appear at the logic design level where there are no pre-defined cells for use in full-custom design. Instead, the designer is free to choose any logic circuit that can be devised for each of the register-level blocks and to divide it up into any types of cell that seem to be suitable. As in semi-custom design, the logic operation is checked by simulation.

The electrical design of the cells requires the use of a circuit simulator to predict the waveforms by solving the circuit and transistor equations. The data for this level of design is provided by the semiconductor manufacturer in the form of parameters for the chosen fabrication process.

In parallel with the top levels of full-custom design, consideration is given to the **floorplan** which divides up the chip area between the main functional blocks. As the shapes and sizes of these becomes more accurately known, the floorplan evolves in parallel with all the main stages in the design. The area for each block contains the layout for the circuit cells which are designed to fit together to cover the entire active area of a full-custom chip. The cell layout is governed by the design rules for the chosen fabrication process.

Layout verification checks that the layout complies with the design rules. Some of the more powerful verification software can also determine whether the layout represents sensible electronic circuits and that it corresponds to the logic netlist of the design. Such checks can be extremely expensive because of the large computer resource required but they can save a great deal of extra cost by locating errors before a design is committed to fabrication.

In the following sections we will consider those stages in full-custom design that have not already been covered in Chapter 8, namely the more flexible logic

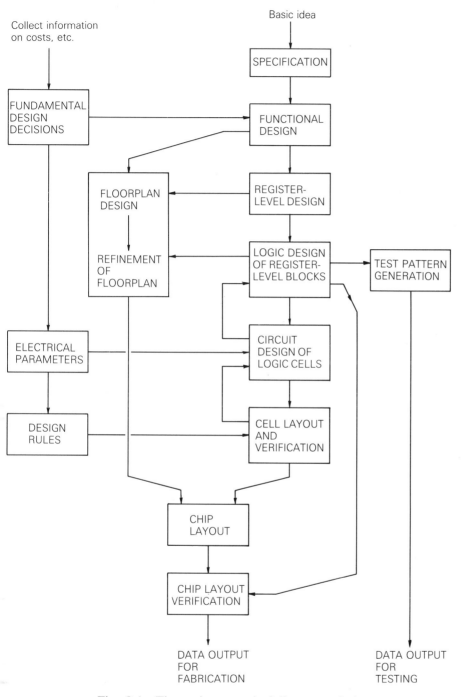

Fig. 9.1   The main stages in full-custom design.

design, and floorplan, circuit, and layout design. A variety of CAD tools may have to be used for these stages because there are very few integrated software packages for full-custom design. This is due to the lack of a large commercial market for full-custom as compared with semi-custom design tools.

Logic Design for Full-custom ICs

In full-custom design, the logic design of register-level blocks such as decoders, adders and counters can use any of the standard logic circuits containing gate and bistable elements that are to be found in textbooks. The logic circuits are themselves composed of cells containing a small group of elements that are chosen by the designer. Care is needed to select the most appropriate logic circuits and cells for a particular application. One aim should be to try to use the smallest possible number of cells for all the blocks in the complete IC because a great deal of time will be spent in designing the circuit and layout for each of them. Full-custom design aims at the repeated and regular use of a small number of cell types. Many logic functions can be implemented using iterative cells rather than random logic and such methods should always be used where possible.

In MOS and CMOS technologies there are some particularly favourable transistor circuits that produce a logic function without the explicit use of individual gates. Examples of these will be given in Chapter 10, but they include the AND–OR–NOT and other complex functions, full adders, multiplexers and circuits using transistors as switches. All these elements have better performance as transistor circuits than if they were composed of gates and they should be used in preference to gates in devising MOS logic circuits. Another preferred element for full-custom design in most technologies is the logic array, or PLA, which is a regular, structured way of generating complex logic functions. The main advantage of PLAs is that they can be designed automatically from a behavioural (Boolean) description, although they are invariably slower than the equivalent random logic.

Other examples are the MOS bistable circuits described in Chapter 4. Although a bistable function can be represented by a combination of gates, the MOS circuit implements it directly in transistors.

The logic design of the register blocks in any style of design must be checked by simulation. As simulation programs may contain only gate and bistable primitive elements, there can be problems in simulating logic circuits that can only be described accurately at the transistor level, such as those described above. In particular, logic simulation of circuits containing MOS switches can give misleading results and it may be better to represent their function in gate or functional form even though there are no actual gates in the circuit. An example of this would be in representing a MOS PLA either by NOR gates or by Boolean equations for simulation purposes.

Another problem in the logic simulation of a full-custom design is that the cell delay values are not known initially; this is unlike semi-custom design where they are given in the library data. The best that can be done is therefore to perform the simulation initially with estimates of the delays based on previous experience (or guesswork!). Once the logic is shown to function correctly, the design of the circuits can proceed, and this will eventually give true delay values that can be fed back for a more accurate logic simulation. This important feedback path, Fig. 9.1, emphasizes the interaction between circuit and logic design in full-custom.

Floorplan Design

As we have seen earlier, full-custom layout contains an enormous number of shapes on many layers. It is virtually impossible to draw such a layout starting randomly with individual circuits and it is essential to adopt a structured top-down approach to the design of layout just as it is for logic. This is the only way that layout can be designed reliably and with the minimum possible effort.

Structured layout design starts by considering the floorplan of a full-custom IC early in the design process when the functional block diagram can still be modified to improve the layout. In designing the floorplan, the objectives are to use the entire chip area for active circuitry and to eliminate long metal interconnections wherever possible. Very little of the area of a full-custom IC should be used for interconnections, unlike a standard-cell or gate-array chip. The freedom to design blocks of any size in full-custom design enables them to be fitted together on the silicon with most of the signal connections made directly across the boundaries between adjacent communicating blocks. There are three advantages in doing this:

1. It increases chip speed by reducing the capacitive loads of the tracks;
2. It reduces the design time by cutting out the laborious and error-prone task of drawing the interconnections at the layout stage; and
3. It usually makes more efficient use of the silicon area.

In mapping the functional block diagram on to the floorplan, attention is therefore given to the flow of data through the chip. If possible, unidirectional flow along a **datapath** is aimed at as in the example of Fig. 9.2 which is a programmable-processor IC with a stack memory. The datapath is horizontal and the control signals for the pipelined processor flow vertically. The floorplan blocks are fitted together into a rectangular chip outline in Fig. 9.2(b) and all the connections are made across the block boundaries with the exception of the data flow to and from the stack. The main circuit blocks can often be seen on the completed chips. Figure 1.1 is a particularly clear example of careful floorplanning.

It must be admitted that IC functions do not always map on to a datapath as easily as in these examples. A pipeline is a very good architecture for full-custom IC layout and the concept is extended to two-dimensional flow in systolic arrays. In general, however, although many chip functions are irregular, they should still be mapped on to blocks that fit together like a jigsaw into a rectangular or square outline. Figure 9.3 shows how **busbars** in (a) and (b) are eliminated by arranging the connection points between A and B to join up when the cells are

A systolic array is a two-dimensional array of simple processing elements, each of which communicates with its four nearest neighbours. This type of architecture is particularly useful for high-speed digital signal processing operations.

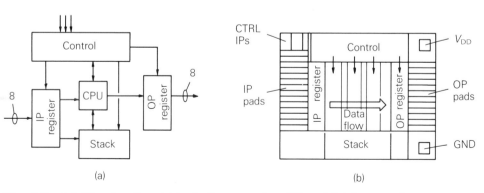

(a) (b)

Fig. 9.2 (a) The functional block diagram of an IC and (b) its mapping on to the preliminary chip floorplan.

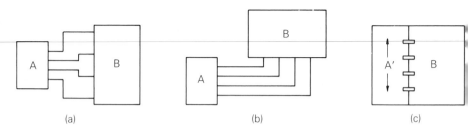

(a)   (b)   (c)

Fig. 9.3   The elimination of busbar connections in (a) or (b) by designing blocks to butt together.

butted together. Although this means that the height of A has to be increased, this is preferable to the zig-zag connections in (a).

The positions of the input and output pads for the bonding of wires to the chip must be included in the floorplan design. Each connection has an associated **pad circuit** for interfacing between the IC and the outside world, and these have to be fitted around the edge of the chip, together with the power-supply pads. The large number of I/O connections typical of large digital ICs usually occupy all four edges, and the chip size might have to be increased to fit them all in. Such a **pad-limited** layout is wasteful of silicon area and it should be avoided by re-designing the chip function to reduce the number of inputs and outputs if at all possible.

Some blocks in the floorplan will eventually contain circuitry that is highly compacted, while others may be expanded to make their connections fit. In designing the preliminary floorplan, some estimates are made of the minimum size of each block. These can be little better than guesses in the absence of previous design experience. However, the approximate area can be calculated if the number of gate equivalents in the block can be estimated. In CMOS, for example, the minimum area of a two-input gate is approximately $600L^2\,\mu\text{m}^2$, where L is the minimum channel length in microns for the process.

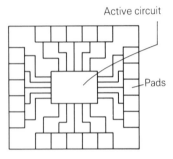

Active circuit

Pads

A pad-limited layout.

Worked example 9.1	The processor block of Fig. 9.2 is estimated to contain the equivalent of 1000 gates. If it is made in a $2.5\,\mu\text{m}$ CMOS technology it will therefore have an area of $1000 \times 600(2.5)^2$ or $3.75 \times 10^6\,\mu\text{m}^2$. Hence a block with proportions as in Fig. 9.2(b) could be $3750\,\mu\text{m}$ wide by $1000\,\mu\text{m}$ high. If the eight input and output pads are spaced at $150\,\mu\text{m}$ it would be worth increasing the height of the processor block to 150×8 or $1200\,\mu\text{m}$ to enable the blocks to butt together. The width might then be reduced to about $3150\,\mu\text{m}$.

The floorplan is gradually refined during the register-level stages of the design by adjusting the sizes and shapes of the blocks as more information on their circuits becomes available. The way of distributing the electrical supply over the chip is also added. Figure 9.4 shows a typical interleaved arrangement of V_{DD} and ground lines for reaching every circuit of the processor chip of Fig. 9.2 from a single pair of supply pads. Complex ICs often use several pairs of supply pads connected together externally to reduce electrical interference and simplify the power supply network on the chip.

In the final layout, each of the floorplan blocks is itself made up in much the

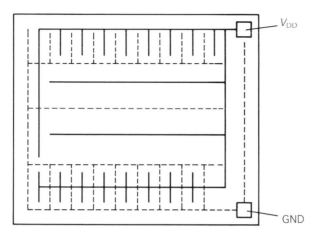

Fig. 9.4 A possible power-supply distribution scheme for the chip of Fig. 9.2(b).

same way by butting together its individual circuits, or **leaf-cells**. The logic design should aim to use the minimum number of different types of cell and to combine them in regular and repeating arrays wherever possible. Regularity ensures reliable and quick cell assembly in design because all CAD layout tools have facilities for combining cells into arrays.

The cell connections are arranged to join up when the cells are butted together in the same way as for the blocks, and irregular connections should still be avoided. As another example of this, Fig. 9.5 shows a situation in which connections are to be made between points in cells A and C without a connection to B. Rather than taking the tracks round the outside of the cells, as in (a), it is better to make all the cells slightly wider so that the signals can be routed right across B as in (b). With present technology there is unlikely to be any electrical interference between the signals in these connections and any underlying circuits in B.

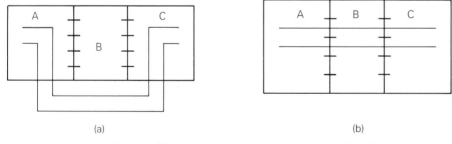

Fig. 9.5 The avoidance of irregular connections between cells (a) by increasing the cell height (b).

Circuit Design and Simulation

There are two stages in designing circuits for ICs:

1. The selection of the best circuit form to use, and
2. The determination of the component parameters to give the best performance.

By 'circuit form' we mean the circuit topology as in a circuit diagram. Examples of some of the most commonly used circuits have been described for various technologies in Chapters 4 and 5 and some further CMOS ones will be given in Chapter 10. Nearly all ICs use such circuits and the designer must be aware of the advantages and disadvantages of any alternatives within the chosen technology.

After choosing the type of circuit to be used, it can be adapted for use in a particular application by the choice of a small number of component parameters. In MOS circuits these are the channel length L and width W of each of the transistors. In bipolar circuits they are the dimensions of the transistor emitters and the resistor values.

L and W are usually whole numbers of microns or half microns and they must be greater than the minimum dimensions that can be fabricated.

In designing digital ICs we are concerned with the propagation delays and power consumption of individual circuits that differ mostly in the electrical load across their outputs. In MOS circuits, the load is purely capacitive and the speed is determined by how quickly it can be charged and discharged. Increasing the width of the transistors reduces their resistance and increases the speed. However, this also increases the input capacitance which slows down the previous circuit so that a compromise has to be reached to give the best overall speed. The larger current taken by wider transistors also increases the power consumption which may then become a limiting factor in n-MOS circuits.

In Chapter 4 it was shown that the delay of a CMOS inverter is approximately $\tau C_L / C_{IN}$, where τ is a characteristic of the process, C_L is the load, and C_{IN} is the input capacitance (Equation 4.11). The same expression can be used for gates in which case C_{IN} is the input capacitance of one of the inputs. Many of the circuits on an IC will have values of C_L / C_{IN} between 3 and 4 and the designer should attempt to eliminate any values that are very much higher than this by adjusting transistor sizes.

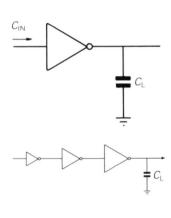

If C_L is very large, as in driving a clock line, two or three inverter stages in series, each with C_L / C_{IN} about 3, will give the lowest overall delay. This is taken even further in designing pad circuits for driving off-chip loads with a capacitance of several picofarads from a minimum sized inverter which may have C_{IN} of a few femtofarads. Six inverters of gradually increasing size would be used to give the lowest overall delay in this case but this might still be the slowest path in the complete IC.

Estimates of circuit performance based on considerations such as these are very approximate because of the many simplifying assumptions made in deriving first-order equations such as (4.11). Such estimates are used in the initial design of an IC but far more accurate equations are needed for refining the design. However, it becomes extremely tedious to evaluate the resulting expressions by hand and it is essential to use a computer simulation that does it automatically.

Circuit Simulation with SPICE

SPICE stands for Simulation Package with Integrated Circuit Emphasis. It originated in the University of California, Berkeley, and was distributed freely until 1980. Since then, commercial versions such as HSPICE, and PSPICE, have improved it in detail and it remains the industry standard for circuit simulation.

There are many computer packages for circuit simulation but the most commonly used ones for IC design are SPICE and its many commercial derivatives. SPICE is a very powerful circuit simulation package which contains built-in mathematical models for integrated circuit transistors, allowing for many physical effects that are ignored in first-order device equations. It simulates the behaviour of a circuit by simultaneously solving the non-linear transistor equations and the time-dependent Kirchoff's law equations for the instantaneous currents and voltages in the circuit. It contains many features, but in digital IC design it is used largely

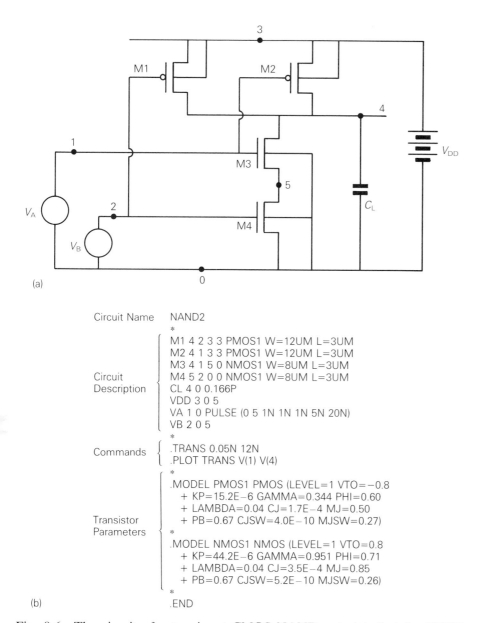

Circuit Name   NAND2
*

Circuit Description
```
M1 4 2 3 3 PMOS1 W=12UM L=3UM
M2 4 1 3 3 PMOS1 W=12UM L=3UM
M3 4 1 5 0 NMOS1 W=8UM L=3UM
M4 5 2 0 0 NMOS1 W=8UM L=3UM
CL 4 0 0.166P
VDD 3 0 5
VA 1 0 PULSE (0 5 1N 1N 1N 5N 20N)
VB 2 0 5
```
*

Commands
```
.TRANS 0.05N 12N
.PLOT TRANS V(1) V(4)
```
*

Transistor Parameters
```
.MODEL PMOS1 PMOS (LEVEL=1 VTO=−0.8
   + KP=15.2E−6 GAMMA=0.344 PHI=0.60
   + LAMBDA=0.04 CJ=1.7E−4 MJ=0.50
   + PB=0.67 CJSW=4.0E−10 MJSW=0.27)
*
.MODEL NMOS1 NMOS (LEVEL=1 VTO=0.8
   + KP=44.2E−6 GAMMA=0.951 PHI=0.71
   + LAMBDA=0.04 CJ=3.5E−4 MJ=0.85
   + PB=0.67 CJSW=5.2E−10 MJSW=0.26)
*
```
(b)   .END

Fig. 9.6 The circuit of a two-input CMOS NAND gate labelled for SPICE (a) with the corresponding SPICE file (b).

to simulate the transient current and voltage waveforms when inputs change state. The graphical output of many versions of SPICE looks very like an oscilloscope trace so that the signal delay, rise- and fall-times (Fig. 4.4) can easily be evaluated.

To use SPICE, a circuit diagram is first drawn and labelled with numbers for the nodes and names for the components, as shown for a CMOS two-input NAND gate in Fig. 9.6(a). The circuit diagram includes the fourth connection to each MOS transistors. This is the **substrate connection**, representing the piece of silicon in which the transistor is made. In a p-well process the substrate connection for the n-channel transistors is made to the p-well, and for the p-MOSTs to the silicon substrate itself. The well and substrate are normally connected to the GND and V_{DD} rails of the circuit as shown. SPICE requires these connections to work out one of the most important second-order effects in MOS transistors which is the change of threshold voltage with source–substrate voltage, known as the **body effect**.

CASS is part of the Silvar-Lisco CAD system.

The circuit data from the labelled diagram is entered into a computer file either by typing text or by using a schematic entry package such as CASS or MINNIE for drawing the circuit on a graphics screen. The SPICE input file produced by either method for the NAND gate example is shown in Fig. 9.6(b). It contains four parts: the circuit name, a list of component connections and values, commands for the output required, and the values of the parameters in the transistor models. Full details of the notation are given in the manuals for particular versions of SPICE but it is worth noting a few of them here:

Node 0 must be the ground or reference node in SPICE. The other nodes are numbered in any order.

1. MOST connections, for example for M1. The node numbers are given in the order drain, gate, source, substrate. The name of the model to be used is PMOS1 and the MOST has the channel dimensions given by W and L (U stands for μm).
2. Output commands. TRANS 0.05N 10N asks for a transient plot with the voltages evaluated every 0.05 ns up to 10 ns. .PLOT TRANS V(1) V(4) asks for transient plots for nodes 1 and 4.
3. Model parameters. These are the values of parameters used in evaluating the characteristics of the MOSTs.

There are different 'levels' of models for MOSTs in SPICE. Level 1, the simplest model, is used for preliminary simulation of a circuit because it uses the least computer time. Level 2 includes many of the additional physical effects in short-channel MOSTs and it uses many more parameters. It is more accurate but takes far longer to run and it is used for the final adjustment of a design. Some of

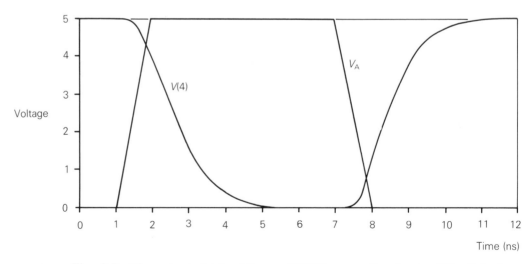

Fig. 9.7 The output obtained from a SPICE run on the circuit of Fig. 9.6 when the *A* input goes from logic 0 to 1 and back to 0 while the *B* input remains at logic 1.

the other levels in commercial versions of SPICE are for models devised by specific companies.

The SPICE model parameters are determined from measurements made on a large number of special test structures. They are provided by the semiconductor manufacturer for the particular fabrication process to be used. The level 1 MOST parameters are described in Appendix 6.

The SPICE program is run with an input file such as Fig. 9.6(b). Figure 9.7 shows the graphical output obtained for this run from which the rising- and falling-edge delays can easily be determined.

Circuit Design using SPICE

SPICE is only used for the electrical simulation of small circuits which might be contained in up to three or four cells on an IC. It gives the delay values that are then fed back into a logic simulator that can readily handle large blocks of circuitry but in far less detail. Logic errors are frequently caused by the generation of glitches due to circuit delays. If they occur, or the logic is not fast enough, SPICE is used to evaluate the effects of changing transistor sizes as suggested by the designer's general understanding of the circuit operation.

To be realistic, the input signals used in SPICE simulations are usually ramp functions as in Fig. 9.7 and they should have a slope that is approximately the same as the output of the previous stage (Appendix 3), which can be found from a few preliminary simulations. The output waveforms often contain kinks and overshoots which are almost certain to be correct because second-order effects make the circuit operation more complicated than expected. Hence, the SPICE results are valuable for finding out whether any of the delays are much greater than anticipated and for making changes if necessary.

There is not much point in refining the circuit design to the point of cutting off fractions of nanoseconds in delays without allowing for manufacturing tolerances which are often very wide. In MOS circuits, the threshold voltage, in particular, may vary by $\pm 30\%$ between chips and SPICE should be used to find what effect this has on circuit delays since the design must still meet the specification in worst-case conditions.

The falling-edge delay here is about 1 ns. The value calculated using the first-order equation (4.12) is 0.8 ns. A level 2 simulation would be more accurate.

The maximum number of transistors in a SPICE simulation is often limited to about 20–30 to prevent the computer time becoming excessive, although the program can handle far larger circuits in principle.

An even more realistic input waveform can be obtained by feeding a ramp signal through a chain of inverters added to the circuit being simulated. Four inverters will round-off the ramp into the typical shape of a signal on the IC.

Two effects that occur in the CMOS two input NAND gate of Fig. 9.6 are due to capacitance at the node between M3 and M4 and the body-effect in M3 since the source of M3 is not grounded. These effects change the output waveform slightly depending on the sequence of input signals.

Cell Layout

As already explained, the layout of an IC is made up of regularly repeating rectangular cells that fit together to cover the entire floorplan. A cell typically contains a circuit such as a gate, a bistable, or a simple combination of these elements.

The design of a cell start with the specification of the relative positions of its external connections on the cell boundary, and the layers on which the connections are made, such as polysilicon or metal. In addition to the connections to the circuit contained in the cell there will often be other tracks that pass right through it as shown in Fig. 9.5. All the connections will later be positioned on a grid with a spacing determined by the design rules for the fabrication process to be used. In CMOS the pitch is at least five times the minimum line width.

Figure 9.8(a) shows the connections of a CMOS cell containing the two-input NAND gate of Fig. 9.6, plus an additional track T_1. The signal inputs A and B

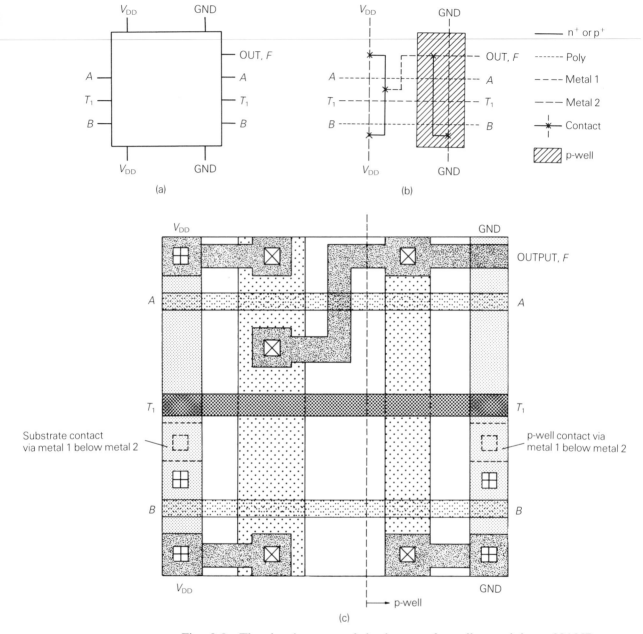

Fig. 9.8 The development of the layout of a cell containing a NAND gate: (a) the relative positions of the external connections, (b) stick diagram, and (c) final layout.

from the left also pass right through the cell for use in further cells on the right. The power supply lines run vertically in this example.

The next stage in cell layout is to devise a **stick diagram** giving the internal topology (Fig. 9.8(b)). The stick diagram shows the layers that will be used for the internal conducting paths in the circuit with different line types or colours for diffusion tracks (n^+ and p^+) and one or two layers of polysilicon and metal. In

Fig. 9.8(b) it is assumed that the process uses a single layer of polysilicon and two of metal. The n^+ tracks are distinguished from p^+ by being inside the p-well which is also indicated on the diagram. The layers are insulated from each other except where contacts are made. Transistors are formed where polysilicon crosses diffusion, making p-channel MOSTs in the p^+ diffusion tracks and n-channel MOSTs in the n^+ tracks.

There are usually several ways of drawing the stick diagram for a circuit and it rarely has the same topology as a conventional circuit diagram. The best stick diagram, found by trial and error, is the one with the minimum number of contacts, which take up a lot of the space on the final layout. The smallest dimensions for the cell are given by the number of grid lines vertically and horizontally in the stick diagram multiplied by the grid pitch.

The final stage in cell layout is to flesh out the stick diagram to the actual layout dimensions which are determined by the design rules that will be considered shortly. Figure 9.8(c) shows the minimum-sized layout for the NAND gate with all the connections on adjacent grid lines. However, to fit with other cells, both vertically and horizontally, the cell might have to be expanded in either direction. Some CAD systems can do this automatically but otherwise the layout has to be redrawn with the connections on the new grid lines.

Quite a lot of space can be saved in assembling cells by sharing power supply lines between neighbours. For example, a cell to the right of our NAND gate can use a common ground line and p-well as shown in Fig. 9.9(a). Design time is saved by using identical cells rotated or mirrored about a vertical or horizontal axis wherever possible. If another instance of the NAND gate does not require the input signals to go right through the cell, the right-hand connections A and B can be left floating as it is scarcely worth modifying the cell by reducing their lengths. To encourage the repeated use of identical cells, bistables are often laid out with both the Q and $\bar{Q}$ outputs even if one of them is not always required.

Some cells are used for routing signals rather than processing them. Figure 9.9(b) shows a cell that might fit on the right-hand side of the previous NAND gate to direct A and B vertically, while F and T_1 are connected to an identical NAND cell. An alternative to adding this cell would be to redesign the original NAND with the outputs is the required places but this might not be worthwhile

It is a pity that this book is not printed in colour which is extremely useful for drawing stick diagrams and layout. The following colours are conventionally used for the different layers:

1. Diffusion – Green
2. Metal – Two shades of blue
3. p-well – Yellow
4. Polysilicon – Red

The transistor is formed where the gate and active layers intersect because the active region is not doped where it is covered by polysilicon in the self-aligned process described in Chapter 3.

Cell layout is started by trial and error. It is easier to produce the first sketches on graph paper and to enter them into the computer only when they are almost correct.

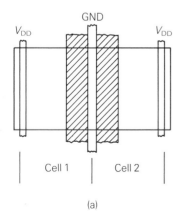

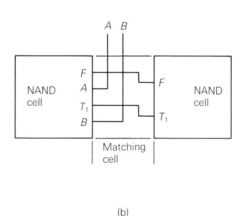

(a)   (b)

Fig. 9.9 (a) Sharing of the ground line and p-well between two cells. Cell 2 is a mirror image of cell 1. (b) An interconnection cell between two NANDs.

if it only occurred occasionally on the chip. At this lowest level cells are sometimes called **tiles** for covering the floorplan.

Layout Design Rules

Many of the layout design rules for a fabrication process express the manufacturer's ability to define shapes reliably in the different layers making up an IC. This depends very much on the particular lithographic and etching processes used. Industrial design rules are extremely detailed for high packing-density processes but simpler **relaxed rules** are usually adequate for ASIC design. Use of these rules might enable a design to be fabricated by more than one manufacturer. In all cases, the design rules form the basis of a contract between the designer and the semiconductor company. If the designer complies by the rules, the manufacturer is duty bound to produce electrically correct ICs.

Design rules apply to the geometry of each layer of an IC and the overlap between layers. For each layer there are rules giving the minimum track width and the spacing between tracks. For a two-micron CMOS process, for example, the polysilicon tracks are at least $2\,\mu m$ wide, and the active region, metal 1 and metal 2, might typically have minimum widths of 3, 4 and $6\,\mu m$, respectively, depending on the resolution obtainable in defining each layer in production. The minimum spacing between the tracks might be $3\,\mu m$ for all the layers.

Metal 2, the outer layer, usually has to be wider than metal 1 because the surface is less flat, which increases etching tolerances.

The design rules for two or more layers reflect the accuracy with which one can be aligned relative to another, either to create or to prevent an overlap occurring. Some examples are given in Fig. 9.10. At (a) the effect of insufficient overlap at x is that the metal does not completely cover the contact window when the shapes are misaligned horizontally, and this would form a high-resistance contact that could lead to a faulty circuit. At (b) the inadequate overlap of the polysilicon and the diffusion edges at x gives a permanently conducting channel between source and drain of what should be a transistor when they are displaced. At (c) the opposite occurs and a transistor with a very peculiarly shaped channel, and hence the wrong characteristics, is formed by the misplaced overlap. Layout with the correct design rules avoids all these faults.

See, for example, Weste and Eshraghian (1985). Layout drawings based on design rules are idealized compared with the final geometry on the silicon. Where sizes approach the technology limit it is not possible to etch sharp corners in layers or contact holes and they may be well rounded off in practice. step edges are also not necessarily vertical or true.

Full sets of design rules are given in more detailed textbooks. The reason for many of the rules can be seen by considering the consequences of breaking them, and this requires some knowledge of fabrication techniques. Others originate in particular electrical effects that are influenced by layout geometry.

The most important of the electrically determined rules for the layout of CMOS circuits arise from the need to prevent **latchup**. This is a fault condition in which the circuit passes a large d.c. current between V_{DD} and ground and fails to respond to input signals. Latchup is due to the fact that the CMOS structure, such as Fig. 2.17, contains parasitic n–p–n and p–n–p bipolar transistors within the silicon. These transistors normally have no effect on the circuit but they can be triggered into a conducting state by a transient voltage spike on the supply or by external radiation, and this causes the MOS circuit to fail. A latched-up CMOS circuit may remain in this state until switched off, when it usually recovers. However, the large current can in some cases destroy the circuit.

The tendency to latchup can be greatly reduced in the detailed design of a CMOS fabrication process. The resulting layout rules usually stipulate a large clearance between the edge of a p- or n-well and the active regions containing MOSTs, and this can waste a lot of area if the layout uses many small wells. To

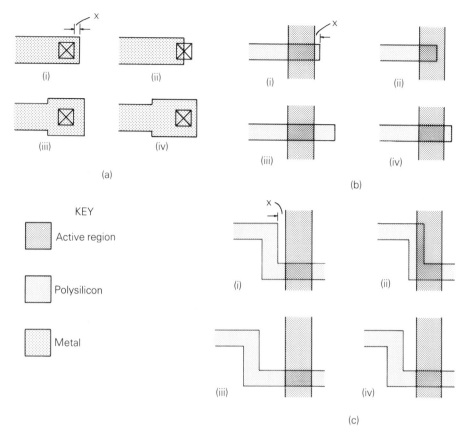

KEY

Active region

Polysilicon

Metal

Fig. 9.10 Examples of two-layer design rules: (a) the metallization over a contact window; and (b), (c) the formation of MOSTs by the overlap of polysilicon and diffusion. In each example four cases are shown: (i) the layers as drawn without adherence to the design rule at x, (ii) the consequence of (i) with the maximum horizontal displacement of the layers, (iii) the layers drawn with the correct design rule at x, and (iv) the consequence of (iii) with the layers displaced horizontally.

avoid this, the design should attempt to group together as many transistors as possible in single wells. This was done in the layout in Fig. 9.8(c) as the well can be extended into neighbouring cells vertically and horizontally to form one large p-well.

The tendency to latchup is also reduced by adding closely spaced contacts between the supply rails and the silicon. For a p-well process, the positive supply is therefore connected to the n-type substrate in every circuit. The ground or negative rail of the supply is similarly connected to the p-type wells that are the substrates for the n-channel transistors. The contact regions to the substrate and well are doped n^+ and p^+, respectively, to make low-resistance connections to the metal. Separate substrate and well connections are included under the metal 2 tracks in the cell layout of Fig. 9.8(c). As an alternative they can be amalgamated in combined contacts between the supply rails and the circuit as in Fig. 9.11.

Latchup is more likely to occur in the large transistors of pad-driver circuits.

The bipolar structures between V_{DD} and ground can be represented by

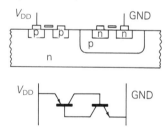

which is very simplified. However, this type of cross connection between p–n–p and n–p–n BJTs always has a high-current condition if the gains of the transistors become sufficiently high. Anything that temporarily increases the current through the circuit may increase the gains sufficiently to trigger latchup.

One of the main advantages of the twin-well CMOS process (p. 26) is that it allows the spacings to be reduced without increasing the tendency to latchup.

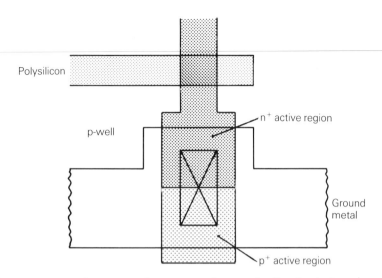

Polysilicon

p-well

n⁺ active region

Ground
metal

p⁺ active region

Fig. 9.11   Layout of a contact from a metal ground rail to both the n^+ source of an MOST and the underlying p-well.

However, it is usually possible for the ASIC designer to use proven, standard designs for input and output pad circuits to avoid the problem.

Layout Capacitances

All the interconnections in an integrated circuit have capacitances which may have an effect on the circuit operation. For full-custom design, the semiconductor manufacturer provides the value of the capacitance per square micron for each of the layers. The layout gives the area of each layer used for connections from which their capacitance can be calculated. These capacitances are then added to the SPICE circuit file to give a more accurate simulation for the particular cell layout, although the effect is often so small that it is scarcely worthwhile if the cell has a reasonably compact layout.

Figure 9.12 shows the capacitances associated with the two-input CMOS NAND gate. The p–n junctions forming the sources and drains of the transistors all have capacitances C_d and C_s which are already included in the SPICE simulations if the junction dimensions are given (Appendix 6). Where the source of a transistor is at a fixed potential its capacitance has no effect on circuit transients, so that C_s can be neglected for all the transistors except M3. The total load capacitance C_L is the sum of the input capacitance of the following stage C_{IN} and the interconnection capacitance C_T. This includes capacitance due to the metal track, a contact, and the polysilicon right up to the gates of the next stage. In one particular design using a three-micron process for a gate with a fanout of three, the capacitances were:

1. Total drain capacitance of gate   28 fF
2. Input capacitance of three minimum-sized inverters   99 fF
3. Interconnection capacitance C_T 36 fF

It can be seen that the input capacitance has by far the largest effect, which would be expected from the small thickness of the gate oxide compared with the other insulating layers.

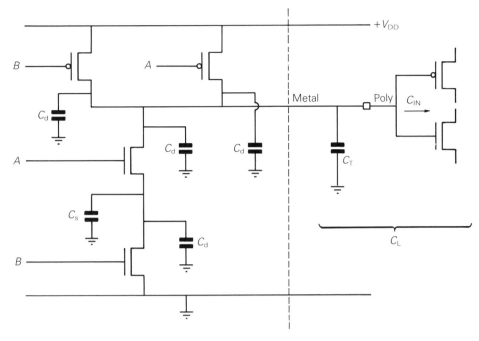

Fig. 9.12   The capacitances associated with a two-input NAND gate. C_d is the drain capacitance, C_s is the source capacitance, and C_T is the track capacitance.

In full-custom design, the interconnection capacitances are always calculated by hand in this way and the CAD tools rarely do anything equivalent to the automatic load extraction from layout that is usual in semi-custom design. This is inevitable with the variety of shapes and layers that are used for connections in a full-custom layout. In general, track capacitances inside cells are small and it is certainly not worth compressing a layout to the final 0.5 µm since the capacitances will only be reduced by a few femtofarads. The capacitance of any connections spanning more than one cell should, however, be included in SPICE simulations.

CAD Tools for Layout

Very powerful CAD tools are required for the design of a complete full-custom chip layout. They need comprehensive graphics editing features capable of handling hundreds of thousands of shapes within a design hierarchy, and links to the tools used for other levels of design for verification. This makes them expensive and not particularly easy to use.

At the level of cell layout, however, a low-cost graphics package running on a PC can be used. It needs the ability to handle shapes on up to 12 superimposed layers that are best displayed in different colours on a high-resolution screen. Full graphics editing facilities are required, such as drawing on a fixed grid, moving, scaling, and merging shapes, and the ability to import repeating pieces of layout such as contacts. A colour plotter is almost essential for checking the layout at leisure. Ideally it should be possible to transfer a completed cell design from a PC to a more powerful system for incorporation in a full-chip design.

A full-chip CAD system needs all these simple graphics editing features plus the ability to combine cells into arrays to form circuit blocks, and to combine small into larger blocks hierarchically. The display should give block outlines or the full detail as required. Some layout systems enable cells to be expanded or compressed within the limits set by the design rules and without changing critical dimensions such as transistor sizes. In some systems, SPICE files for simulation can be produced automatically from cell layout, while others enable cells to be laid out automatically from stick diagrams. The output of a full-chip CAD system is a file written in one of the standard formats for chip layout such as the electronic design interchange format (EDIF) which can be used directly for making masks.

Design-rule checking (DRC) is an essential feature of all layout systems. In **on-line DRC**, the layout is checked while it is being drawn and errors are immediately shown up. In other systems, DRC is run at the designer's request but it is always worth checking that cells are correct before combining them into larger blocks as it becomes more and more difficult to correct errors as the layout expands. The positions of design rule violations are superimposed on the display of the layout and a text error file is also usually produced.

The layout packages commonly used in educational institutions in the UK (Silvar Lisco's Princess and Racal-Redac's Isis) contain many of the above features. Isis, in common with many other modern layout tools, can generate layout from a symbolic input similar to a stick diagram. With automatic cell compaction this can reduce the layout design time by a factor of four compared with a basic polygon editor.

Software for checking layout beyond DRC is less commonly available, although verification packages are produced for checking (a) that a layout makes sense electrically in that all nodes have paths to power supply lines and that there are no short circuits, and (b) that a layout agrees with a transistor-level circuit diagram. Such packages require very large computer resources and they are only used in major industrial design centres.

Summary

The term 'full-custom' is used in this chapter for design that goes down to the circuit and layout levels for the entire chip without the use of any predefined parts. Advanced semi-custom methods are now far more reliable and this type of full-custom design is rarely, if ever, undertaken except as an educational exercise. However it is important to understand the full-custom methods for the design of cell libraries and for small full-custom circuits to be added to designs built up from previously used parts or semi-custom libraries.

Full-custom design includes all the top-level design stages of semi-custom design, with the addition of floorplan, circuit and cell-layout design. Circuits can be optimized by fully understanding their electrical operation and by checking the performance with SPICE simulations. Circuit topology is converted to layout via stick diagrams. The layout is constrained by design rules that allow for the effects of geometrical tolerances in fabrication.

There are many opportunities for making slips in full-custom design and it is very difficult to verify that the layout of a full-custom chip is absolutely correct before fabrication.

CMOS Circuits for Full-custom and Analogue ICs   10

Objectives

Studying this chapter will enable you
- [ ] To learn more about static CMOS logic circuits using transistors as switches rather than NAND and NOR gates. The multiplexer and AND–OR–NOT functions become preferred elements in full-custom logic design.
- [ ] To understand the principles of dynamic CMOS circuits such as domino logic and their problems and advantages.
- [ ] To become aware of the uses for mixed analogue–digital ASICs and of the semi-custom methods of designing them that are the same as for purely digital ICs.
- [ ] To appreciate the uses for purely analogue CMOS ICs and, by considering the circuit of a simple CMOS operational amplifier, to find out that analogue design of circuits made entirely of n- and p-channel MOSTs is very similar to bipolar design.

The basic gate and bistable circuits used in CMOS digital ICs were described in Chapter 4. All combinational logic functions can be built up from such simple inverting gates, but in MOS, unlike bipolar technologies, this is not necessarily the best way of designing digital systems because simpler, transistor-level circuits can be used to provide alternative logic functions. These circuits are both smaller and faster than gates, and their availability gives a different emphasis to logic design. The two classes of CMOS circuit, static and dynamic, are briefly described in this chapter. Both types can be used in full-custom design. However, dynamic circuits are not often used in semi-custom design because they are less robust as general-purpose building blocks.

We consider CMOS circuits only in this chapter because it is the technology likely to be available to most readers for fabrication.

Static Integrated-logic Circuits

Complex Gates

Complex gates are circuits which evaluate combinational functions directly rather than through a network of NAND and NOR gates. An example is shown in Fig. 10.1(a) where the function F is produced by a complex gate containing eight transistors. With inverting gates (Fig. 10.1(b)) 14 transistors would be required to produce the same function. In general, a complex gate contains p- and n-channel **evaluation blocks** as shown in Fig. 10.2. In the n-block, a transistor is switched into a conducting state by a logic 1 at its input. The circuit is designed so that certain combinations of 1s form conducting paths through the block from output to ground, giving a logic 0 output. With all other combinations the design ensures that there is a conducting path from the output to V_{DD} through the p-channel MOSTs that are turned on by logic 0 inputs. The transistor connections

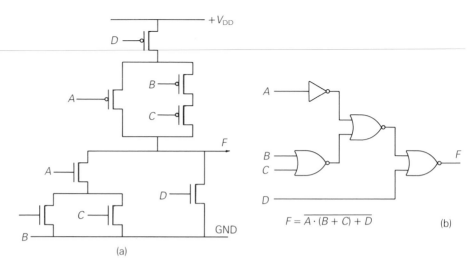

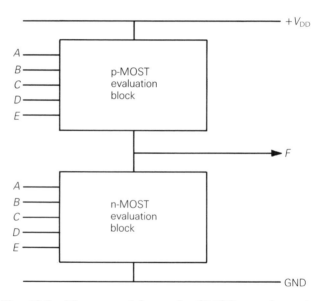

$$F = \overline{A \cdot (B + C) + D}$$ (b)

(a)

Fig. 10.1   A CMOS complex gate circuit (a), and its NOR equivalent (b).

Fig. 10.2   The general form of a CMOS complex gate.

can therefore be arranged to define any logic function, F. It can be seen from Fig. 10.1(a) that series connections in the n-channel evaluation block become parallel connections in the design of the p-block and vice versa.

When the output of a complex gate changes state the delay time depends on the total resistance of the p-block for pulling up, or the n-block for pulling down the output. To maintain the speed the transistor sizes in complex gates therefore have to be adjusted to give the same total resistances as in gate circuits on the same chip. If W_n and W_p are the widths of the n- and p-channel MOSTs of a minimum sized inverter elsewhere in the circuit, the widths of the transistors in Fig. 10.1(a) would therefore be as in Table 10.1. These widths ensure that the resistances are approximately the same as in the inverter in the worst case where

Table 10.1

n-channel transistor width			p-channel transistor width	
D	W_n		D	$2W_p$
A	$2W_n$		A	$2W_p$
B	$2W_n$		B	$4W_p$
C	$2W_n$		C	$4W_p$

only one path is switched on through either the n- or p-channel block. When more than one path is conducting the resistance is reduced further and the smaller delay that results is a bonus.

Design an exclusive NOR (XNOR) circuit using a complex gate assuming that the inputs A and B, and their inverses are available. Can the circuit of the p-block be simplified?   **Exercise 10.1**

A disadvantage of complex gates is that their layout can become very irregular, both in the circuit topology and the transistor sizes, as shown by the example above. As a result the layout capacitances increase with the complexity of the function so that complex gates need to be used with care. An alternative approach is to change the logic in order to use circuits that can be constructed from regular chains of p- and n-channel transistors.

Regular-layout Circuits

A circuit containing a chain of two p- and two n-channel transistors is shown in Fig. 10.3(a). It is a **tristate inverter** that is exactly the same as a normal inverter if the select input S is 1, but it has a high impedance output if $S = 0$. The outputs of two tristate inverters can be wired together if they have opposite S inputs since only one is active at any time. The circuit so formed (Fig. 10.3(b)) is an inverting multiplexer made with only eight transistors compared with the 14 needed for a gate implementation of the same function.

A closely related circuit, giving the AND–OR–NOT function of four variables is shown in Fig. 10.4. If the link between the two chains is placed between the pairs of n- rather than p-channel transistors the function is changed to OR–AND–NOT. Alternatively, if the input C is replaced by A, and D by B, the original circuit gives the exclusive OR function of A and B as in Exercise 10.1. These circuits therefore have many applications.

The advantages of regular layout in full-custom design are so great that the logic design should use multiplexer and AND–OR–NOT functions wherever possible. Bistables can also be fitted into the same regular layout by using the inverting multiplexer of Fig. 10.3(b) with feedback, as shown in Fig. 10.5. This is actually the same as the transmission-gate D-type described in Chapter 4 (Fig. 4.21(c)).

With the appropriate transistor sizes, all these circuits can be designed to operate at about the same speed as the smaller static gates. Even simpler circuits

In logic design with standard ICs one used to minimize the number of gates. In CMOS circuits it might appear that it is the number of transistors that should be minimized. However, the transistors occupy a very small part of the layout and a better criterion might be to minimize the number of contacts in a circuit.

163

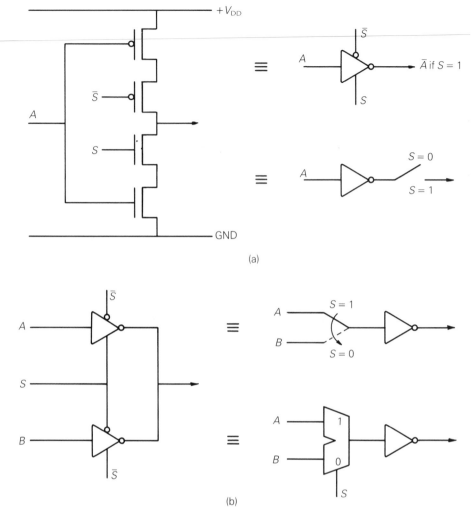

Fig. 10.3 A tristate inverter circuit (a), and its use for making a two-input inverting multiplexer (b).

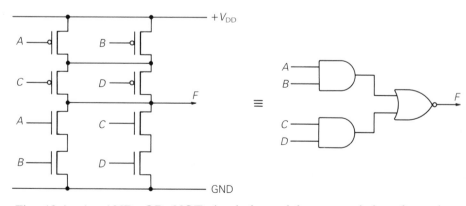

Fig. 10.4 An AND–OR–NOT circuit formed from two chains of transistor pairs.

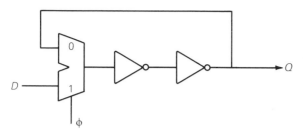

Fig. 10.5   A transparent D-type constructed from the circuit of Fig. 10.3(b).

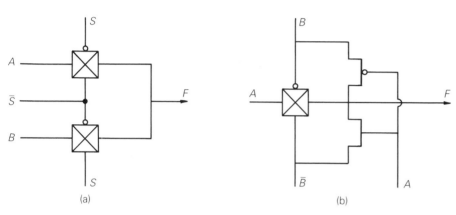

Fig. 10.6   Two switch-level CMOS circuits: (a) a two-input multiplexer with four transistors, (b) an XOR with six transistors including two in the inverter for *B* (not shown).

can be made using transmission gates or single transistors in switching circuits energized from the signal sources rather than the d.c. supply. Two examples are shown in Fig. 10.6 and others may be found in more detailed books. If any of these is used in full-custom design, its function should be simulated in all conditions to ensure that it works correctly and at the required speed for its particular application.

Dynamic Logic Circuits

In all static circuits, the logic state, 1 or 0, at the output of a gate is determined by a conducting path to either V_{DD} or ground. With MOS technologies it is also possible to make dynamic circuits in which the logic state is represented by the charge stored on the capacitance of a node that is regularly disconnected from the supply. There are many types of dynamic circuit but only the most common ones will be described here. They all require a clock signal for their operation so that they are inherently suitable for use in synchronous digital ICs where they have substantial advantages over static circuits.

   Figure 10.7(a) is the circuit of a **precharged dynamic gate**. It is far smaller than a static complex gate because it uses only a single evaluation block containing the better n-channel transistors. Its operation over one clock cycle is shown at (b). When $\phi = 0$, transistor M1 is on and the output capacitance C_L is precharged

Dynamic circuits are not made in bipolar technologies because of the greater leakage current through bipolar transistors in the OFF state. This would rapidly discharge any isolated capacitor.

A complex gate is inefficient because every input switches both an n- and a p-channel transistor. This represents logical redundancy. The input capacitance of the circuit in Fig. 10.7(a) is less than half of that of a complex gate, which is one of its main advantages.

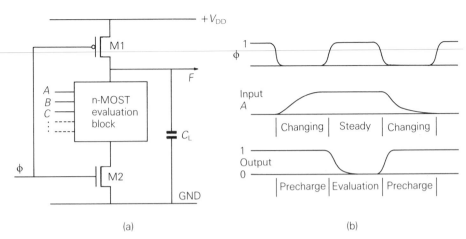

(a)   (b)

Fig. 10.7   A precharged dynamic CMOS logic circuit (a), and its timing diagram (b).

Problems also arise due to **charge sharing** in dynamic circuits. This occurs when a switch connects an isolated capacitor to another circuit node. The charge may then be shared with the additional capacitance causing the voltage to fall to some poorly defined level between V_{DD} and ground.

Two more subtle reasons for adding the inverter are:

1. That without it the delay in discharging the precharge voltage when ϕ first becomes 1 may inadvertently discharge the following stage. With the inverter, the output is at 0 during and just beyond the precharge period and it can have no deleterious effect on the following stage.
2. That it eliminates charge sharing problems when the transmission gate closes.

to V_{DD}, logic 1. The signal input voltage is allowed to change while $\phi = 0$ but it must remain constant for the next half cycle. When $\phi = 1$, M2 is like a closed switch and M1 is open. The output is then discharged if there is a conducting path through the evaluation block due to its logic function being satisfied. The output only becomes valid during the $\phi = 1$ half cycle, and it is always in the precharged logic 1 state during $\phi = 0$.

In precharge circuits of this form, there are problems in connecting one stage to the next due to the constraints in the signal timing. These are overcome by using different clock signals for adjacent circuits, and a common arrangement is to use a four-phase clock, generated on the chip, with dynamic circuits.

Successive stages of dynamic logic, all working on the same phase, can, however, be connected in cascade, with the simple addition of a static inverter in each stage, as shown in Fig. 10.8. In this arrangement, the evaluation of the first stage when appropriate input signals are applied makes X go low and Y high. This switches on a transistor in the second stage and it can lead immediately to its discharge if the other inputs are also high. If the multi-stage logic function of the entire circuit is true, successive stages therefore discharge rapidly one after the other during one half clock cycle.

The name of this circuit, **domino logic**, comes from analogy with the collapse of a row of dominoes when the end one is pushed over. The transmission gate at

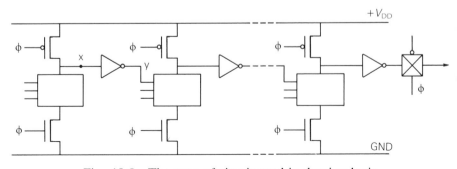

Fig. 10.8   The type of circuit used in domino logic.

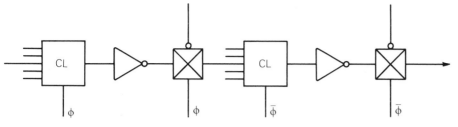

Fig. 10.9 The use of domino logic in part of a pipelined processor.

the final output of the circuit is closed during the evaluation time so that the output can be passed on to another circuit which must operate in antiphase with the first. Figure 10.9 shows how an arrangement of this type can form part of a pipeline processor, functioning like a shift register but with the addition of logic at each stage.

Domino logic enables a large logic function to be evaluated in a very short half clock cycle and it is far faster than the corresponding static complex gate. However, all dynamic circuits require careful electrical and layout design and thorough simulation to ensure reliable operation in a particular IC, and this explains why they normally are only used in full-custom designs.

There are usually far more of the higher mobility n-channel MOSTs than p-channel MOSTs in dynamic circuits. The processing is therefore optimized by using an n-well process with the n-MOSTs fabricated in the substrate.

Semi-custom Analogue–Digital ASICs

Although the emphasis throughout this book has been on digital ICs, it would be incomplete without relating it to analogue design. Readers will be familiar with standard-product analogue ICs, from op amps upwards, that are widely used in electronics. Nearly all these ICs are bipolar because this technology is fast and it provides the high drive capability required for general-purpose components.

There are many needs for ASICs that have analogue and digital circuits on the same chip. They are frequently required for instrumentation and control applications where the chip input is an analogue signal from a sensor and the output is a control signal derived from digital processing on the chip. The analogue input may be amplified, switched, sampled, compared with a reference voltage, or converted to digital form by an analogue to digital (A/D) converter. The output may be a single ON/OFF signal or an analogue voltage derived from a digital to analogue (D/A) converter on the chip.

Mixed analogue–digital applications often require a far higher output voltage or current than would be normal in purely digital ICs. The fabrication processes are therefore often adapted to make high-voltage or high-current transistors and perhaps special components such as improved resistors and capacitors in chips that include analogue functions.

Many semiconductor companies produce semi-custom ASIC products with mixed analogue–digital capability. They are both standard-cell, fully masked ASICs, and the equivalent of gate arrays that can be called **linear/gate arrays**. They are produced in both bipolar and CMOS technologies.

The design of a semi-custom analogue–digital IC uses predefined analogue circuits provided in the cell library together with the normal digital functions. Typical analogue cells are op amps, comparators, A/D and D/A converters, oscillators, voltage reference sources, and output drivers. The library contains data on the electrical performance and size of the cells. The internal layout of the circuits is added by the semiconductor manufacturer when the masks are made.

Analogue circuits need transistors of various sizes, and, in bipolar circuits, resistors of a range of values. In linear/gate arrays different components are therefore provided in the areas to be used for analogue and digital cells. The digital elements are usually grouped into the centre of the chip and the analogue and I/O cells are around the outside. The cells are laid out and interconnected in these areas in exactly the same way as for purely digital designs.

There are problems in adequately simulating and testing mixed analogue–digital ICs that have not yet been completely solved. Although both parts can be treated separately, this does not allow for all possible interactions between them such as digital transients affecting the analogue functions. Improved CAD tools are therefore being developed for the simulation of larger mixed circuits where the interactions cannot be so easily predicted by the human designer.

Full-custom Analogue CMOS Design

Most analogue circuits are based on the ability to amplify signals more or less linearly. At first sight, MOS transistors appear to be greatly inferior to bipolar ones for this because their control action is far weaker (Fig. 5.20(b)) and this reduces the amount of amplification that can easily be obtained. However, MOS, and particularly CMOS, circuits have the great advantage of low power consumption that is just as important for complex analogue ICs as it is for digital ones. Although CMOS operational amplifiers, for example, usually have far lower gain than bipolar ones, it is possible to put a large number of them on a single chip for implementing complex signal processing functions. Many of the applications of CMOS analogue ICs are therefore for making high-performance filters and other circuits for use in communications systems.

An analogue signal carries information in its waveform. Accuracy in processing the information requires that the response characteristics of circuits should be precisely defined even though their components suffer from non-linearity, poor tolerances, and temperature dependence. The art of analogue design has always been in making good circuits out of bad components. This is even more of a problem for ICs where small components have particularly poor characteristics. The problem does not occur to the same extent in digital design where the voltages and currents do not need to be so accurately defined.

Full-custom methods are used in analogue IC design to get the highest performance in complex systems, which is even more important than for digital circuits. Although semi-custom design is quick, it loses performance because predefined circuits are not optimized for use in any particular application.

As an introduction to full-custom analogue design we will consider the basic circuits used in CMOS operational amplifiers. They are developed from MOS implementations of the more familiar bipolar circuits.

Resistors for CMOS ICs

High value resistances of tens or hundreds of kilohms are needed for analogue MOS circuits. These values are too high for fabrication by the normal doping procedures used for making the resistors in bipolar ICs. Instead, the resistors are

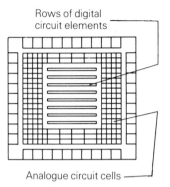

Rows of digital circuit elements

Analogue circuit cells

See, for example, G.J. Ritchie (1987) for the basic bipolar circuits used in op amps.

168

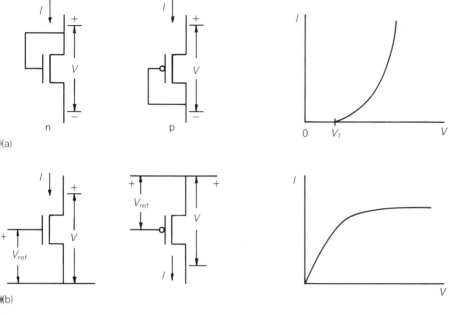

Fig. 10.10 Two ways, (a) and (b), of connecting MOSTs to make resistors, and their resulting I–V characteristics.

usually made from MOS transistor structures fabricated by standard CMOS processes. Although these resistors have very non-linear characteristics this is not necessarily a disadvantage. If extra processing can be allowed, more precise and linear resistors can be made for analogue ICs from specially doped polysilicon tracks, but at additional expense.

Figure 10.10 shows how both n- and p-channel MOSTs can be connected to give the two-terminal conduction characteristics shown. In (a), the gate is connected permanently to the drain of transistor. To find the I–V relationship we need to identify points on the transistor characteristics where $V_{GS} = V_{DS}$. Recalling that the MOS transistor becomes saturated when $V_{DS} = (V_{DS} - V_T)$ (Fig. 4.23) these points can be seen to lie on a curve parallel to the saturation curve but displaced to the right by a distance V_T. In this condition the transistor is therefore always saturated and the expression for the upwards-curving characteristic of the resistor is

$$I = \frac{\beta}{2}(V - V_T)^2,$$

where β and V_T have the values appropriate to the particular type of MOST. The current at any voltage can be adjusted in design by changing the value of β which is proportional to the channel aspect ratio W/L for the structure.

Another connection used to make resistors is shown in Fig. 10.10(b). The fixed potential V_{ref}, obtained from elsewhere in the circuit, is applied between source and drain, and the I–V curve is just the characteristic curve of the transistor with this value of V_{GS}. The almost constant current that flows through this resistor is again determined by the choice of the W/L ratio in the design of the circuit.

Remember that the drain is defined as the most positive terminal of an n-channel MOST and the most negative for a p-channel device.

169

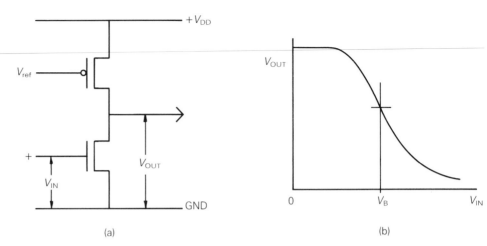

(a)   (b)

Fig. 10.11   A simple CMOS amplifier and its transfer characteristic.

CMOS Amplifier Circuits

A simple inverting amplifier is made by connecting a MOS transistor in series with a load resistor. The amplifying device is usually an n-channel MOST and the resistor can be any of the types described above. Figure 10.11(a) shows an amplifier with a constant-current load resistor.

This circuit can be analysed graphically, in exactly the same way as for a bipolar amplifier, by superimposing the load-line characteristic on the output current–voltage curves of the amplifying device. The resulting form of the transfer characteristic is shown at (b). It is fairly linear for a small range of input voltages on either side of a d.c. bias voltage V_B that would be added to a signal in using the circuit as an a.c. amplifier. The shape of the curve depends on the W/L ratios of both transistors that are determined in the design of an amplifier to give a specified gain and input voltage swing. The circuit takes a steady d.c. current which is also determined by the design.

A CMOS differential amplifier is shown in Fig. 10.12. If is fundamentally

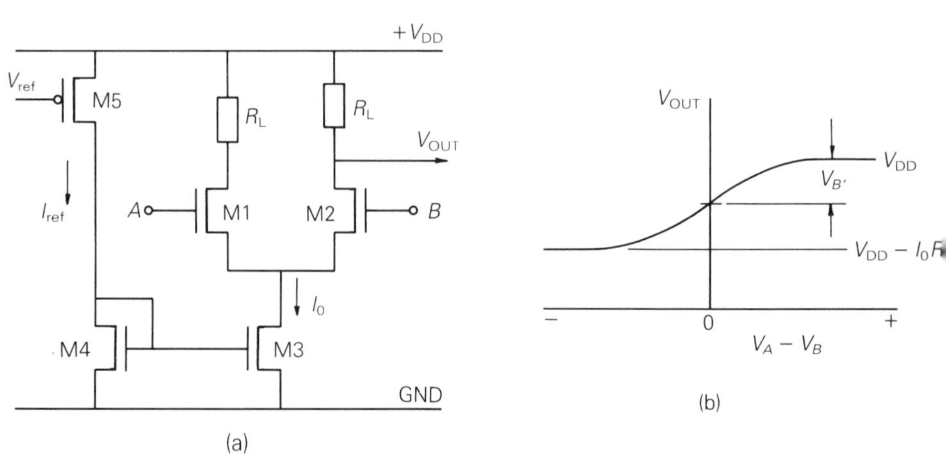

If the gate of the p-channel MOST in Fig. 10.11(a) was connected to the input rather than to V_{ref} the circuit would be the same as a CMOS digital inverter. As an amplifier, an inverter has a higher gain over a smaller input voltage range than the one described.

Fig. 10.12   A simplified CMOS differential amplifier circuit and its transfer characteristic.

170

the same as the bipolar differential amplifier that forms the basis of ECL logic (Fig. 5.13). The amplifying transistors are n-channel MOSTs M1 and M2, and the almost constant current I_0 is controlled by the current-mirror circuit M3, M4. The reference current I_{ref} is determined by the constant-current resistor M5 and its reference voltage. The circuit is shown with a single-sided output from the drain of M2. The d.c. output voltage is $V_{B'}$ below the positive supply V_{DD} when the circuit is balanced with zero differential input voltage. This output can be used as the bias voltage of a next stage of amplification.

See G.J. Ritchie (1987) for an explanation of the bipolar current mirror circuit.

A Simple CMOS Operational Amplifier

The two basic amplifier circuits described above are slightly modified and combined to make the very simple CMOS operational amplifier shown in Fig. 10.13. It is drawn with the positive and negative supplies $+V_{\text{DD}}$ and $-V_{\text{SS}}$ that are normally used in MOS analogue circuits. The main change compared with Fig. 10.12 is that the differential stage is turned upside down by replacing the n-channel transistors M1, M2, and M3 with p-channel devices, and also reversing the polarities of transistors M4 and M5 in the potential divider. A third change is that the load resistors R_{L} are replaced by an active load formed by the current mirror M6, M7, which behaves like a very high resistance.

The reason for inverting the differential circuit is to enable it to be directly connected to the second stage, with the d.c. output $V_{B'}$ providing the bias for the single-ended amplifier. The current I_1 is proportional to I_0 as it is determined by the same current mirror, M4, M3, M9. The **compensating capacitor** C is added to increase the phase shift in the second stage to ensure the a.c. stability of the circuit.

An operational amplifier of this type can have a gain of a few thousand, a power consumption of less than a milliwatt, and a size of about $300 \times 450\,\mu\text{m}$ using three-micron design rules. There are many more complicated circuits that can be used to improve particular aspects of its performance.

These examples of simple amplifiers have been given to show that the methods used in designing in CMOS are the same as for any other circuit technology. CMOS circuits are, in fact, easier to design than bipolar because the transistors take no input current and their first-order equations are even simpler.

Another advantage of changing from n-channel to p-channel transistors, M1, M2, for the input stage is that p-channel MOSTs have lower low-frequency noise figures.

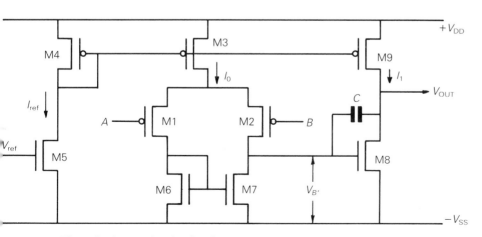

Fig. 10.13   A simple CMOS operational amplifier circuit.

171

The Analogue Design Sequence

The design of analogue CMOS circuits is started by choosing a suitable circuit that might be able to meet the specification, usually from books or journal articles.

See, for example, Allen and Holberg (1987).

Analysis of the circuit by hand using first-order transistor equations can show up the main trade-offs between various performance parameters, for example, the gain of an amplifier against power consumption and frequency response. The only design variables for a particular fabrication process are the width and length of the transistor channels. The values of all the other electrical parameters will have been obtained from the anticipated manufacturer.

Analogue design is far more open than digital because there are many more analogue parameters of a circuit, any of which might be crucial for a particular application. For the operational amplifier, for example, in addition to the gain and frequency parameters, any of the following might have to meet a particular specification: noise, distortion, maximum input or output voltage swing, common-mode rejection ratio, power-supply rejection ratio, input offset voltage, and output slew rate. The factors determining any of these can be evaluated in the preliminary design. Many of them require trade-offs between one aspect of the performance and another. Some may only be met by using a different form of circuit.

The preliminary hand calculations give first-order estimates of the values of L and W that might be used for each transistor. The performance is then evaluated more accurately using SPICE simulations and it is improved by adjusting the dimensions in ways suggested by the designer's understanding of the circuit operation. It is important to allow for tolerances in the parameters and temperature variations so that the required performance is obtained in worst-case conditions.

The layout of an analogue circuit is done in exactly the same way as for a full-custom digital cell, using the same CAD tools. It is, however, more important to evaluate the parasitic capacitances from the layout and feed them back into a final SPICE simulation if the frequency response of the circuit is important.

Summary

This chapter has extended the description of CMOS circuit design (in Chapter 4) to introduce some additional digital circuit forms that can be used in full-custom design. It is shown that CMOS circuits can often use chains containing two n-channel and two p-channel MOSTs connected in various ways and this is an advantage for regular layout. Dynamic circuit forms contain fewer transistors and they are generally faster than their static equivalents but they need careful electrical design.

Analogue functions are often included in cell libraries for mixed analogue–digital semi-custom design. Analogue CMOS circuits have similar topologies to their bipolar equivalents but with the resistors made from diode-connected MOS transistors. Circuits such as operational amplifiers are far smaller in CMOS and they take very little power but their other properties are not as good as their bipolar equivalents. Their design requires a good understanding of traditional electronics circuit design that is outside the scope of this book. Designs are checked by using SPICE and laid out in much the same way as for digital cells.

VLSI Design and Future Trends

11

Reading this chapter will enable you
- ☐ To understand the differences between this introductory view of IC design and the techniques used in industry for VLSI design.
- ☐ To realize the rate at which IC fabrication technologies are developing and the consequences in the near future.

Objectives

This book is only an introduction to IC design and technology and we should conclude it by considering briefly how the subject is extending in two directions: to the increasing complexity of VLSI design, and to the future.

The Problems of VLSI Design

In Chapter 1 we defined VLSI as a circuit complexity of more than 10 000 gate equivalents going up towards hundreds of thousands or even millions of gates. This is an immense increase in complexity and far beyond what most readers of this book are ever likely to design.

There is no problem in designing VLSI ICs covering large areas of silicon with simple repetitive functions. The problems arise when VLSI is used to its full capability to make highly complex systems such as 32-bit digital signal processing ICs or advanced microprocessors. The scale of the design task at both the systems and the circuit and layout levels is then increased by orders of magnitude and it can only be undertaken by a whole team of engineers working, for perhaps several years, on the design of a single IC. As in any large-scale engineering project, this raises problems of design management and coordination.

Since the design of a complex VLSI chip requires a large investment, it is only undertaken in industry when a large potential market is foreseen, and this has many consequences for production and design. The final chips will almost certainly be fabricated using a full set of masks to get competitive performance and the smallest possible chip size for mass production. This rules out the use of gate arrays which, although available with up to 100 000 gate equivalents in channelless arrays, are inefficient for the mass production of VLSI functions. Of the other approaches, the simple standard-cell method of making ASICs with routing channels as described in Chapter 6, similarly wastes too much chip area to be efficient for VLSI. Finally, the full-custom hand-crafted design of complete VLSI chips is also out of the question as it is far too labour intensive and unreliable. Different approaches to layout design are therefore required for highly complex chips. These increasingly use the automatic generation of mask patterns for large circuit blocks.

The *principles* of design described in this book are even more essential for VLSI, in particular the need for hierarchy in developing both the functional

173

design and the floorplan. At the bottom of the hierarchy the layout of regular circuit blocks composed of repeating cells is, however, well suited to automatic generation in which large blocks of circuitry are compiled in the CAD system by assembling many simple predefined parts. The designer requests a block of a certain functional size with certain options and the layout and a simulation model are produced by the software. This style of design with parameterizable cells was described at the end of Chapter 8.

Automatic block generation can be taken further by allowing the basic leaf-cells to be stretched so that they butt together for making connections without additional routing tracks. The assembly of such blocks into the floorplan is easier for the designer because their exact minimum sizes are known and routing can be done automatically. A few full-custom cells may be added for the less regular functions that cannot be generated automatically and for peripheral cells round the blocks. Many VLSI ICs are designed by this or similar methods, right up to the maximum complexity. Although based on standard-cell parts, they use the silicon area nearly as efficiently as in a hand-crafted full-custom design.

Improved CAD tools will eventually reduce the design time for VLSI even further, enabling it to be used to its full potential for low-volume chips. The development of tools for compiling larger and more varied blocks is one step in this direction. Further ahead, the ultimate aim of some of the long-term research in CAD is the production of a full-chip **silicon compiler** that would produce a complete layout, simulation results, and test vectors from the chip specification. This is a far larger task than compiling blocks because it replaces the designer for the most creative level which is in choosing the best strategy for producing the functional behaviour required. Until true silicon compilers are in regular use, there will still be a lot of satisfying work for VLSI designers.

Several CAD companies already produce software packages that they call **silicon compilers**. These mostly generate circuit blocks but not complete chip layouts. A true silicon compiler is a long way off.

Technology Trends

It is usually unwise to make predictions about the future. However, it is comparatively safe to forecast trends in silicon IC technology because of the large amount of successful research and development already done on reducing minimum feature sizes down to $0.5\,\mu m$. This is certain gradually to seep through to production giving the trend already described in Chapter 1, page 7).

The incentive for the semiconductor industry to reduce the minimum dimensions in ICs is largely economic. Halving the dimensions quadruples the packing density on a chip which either reduces the cost or increases the complexity for the same cost. Another advantage is that delays are reduced as circuits are scaled down. Equation (4.10) showed that for CMOS the propagation delay of an inverter is

$$t_{pd} = \tau C_L / C_{IN}.$$

If the entire circuit is scaled down, C_L / C_{IN} is unaffected, but τ is reduced, as it is proportional to L^2. The improvement is less than this in practice because not all circuit dimensions are scaled down equally. A limit to the maximum frequency of operation of digital ICs is eventually set by delays in signal propagation along the interconnection tracks that become more important as dimensions are reduced.

In particular, the sizes of contacts and metal tracks, which take up a large proportion of the chip area, are not necessarily reduced proportionally to the channel length dimension.

A limit to the smallest size of conventional circuits is finally determined by power dissipation, even for CMOS. The power dissipation of a CMOS inverter changing state at a frequency f was shown in Chapter 4 to be approximately

$$P = C_L V_{DD}^2 f.$$

If the circuit is scaled down, C_L is reduced, but, if the opportunity is taken to increase f, the power per circuit remains constant. However, the number of circuits on a chip is increased as they are scaled down so that the total power rises. This will cause increasing problems in ICs fabricated with sub-micron dimensions and the only way round it is to reduce V_{DD} which seems certain to happen. The power dissipation of bipolar circuits usually limits the packing density even with present technology.

The rise in chip complexity brought about by reduced dimensions will be matched by improvements in CAD systems as outlined above. However, it will also aggravate the problem of testing VLSI chips in production quantities. This will be helped by increasing the use of self-testing circuits in all the main functional blocks of ICs since the extra area required for built-in self-test becomes less significant as circuit dimensions are reduced.

The likely increase in the maximum chip size (page 7) is determined by yield and commercial factors. A dramatic increase could occur with **wafer-scale integration** in which circuits will test and reconfigure themselves to avoid defects, not necessarily over an entire wafer, but certainly in far larger areas of silicon than present-day chips.

Increases in complexity will have less effect on the small, low-frequency, ASICs that account for the bulk of present gate-array production. However, the rapid advances in programmable ICs towards higher speeds, greater complexity, and more flexible architectures, will increasingly erode the market for low-volume masked ASICs. The production quantities at which one technology becomes more economical than another, as shown in Fig. 6.12, will need continuous monitoring as the technology continues to develop. The larger gate-array and standard-cell ICs will increasingly contain blocks of dedicated circuitry such as microprocessors, ROM and RAM which will further blur the distinction between semi- and full-custom design in industry.

The advantages of reducing dimensions are far less obvious for analogue functions. The electrical characteristics of MOS transistors deteriorate as they are scaled down and their tolerances get worse. It is not therefore clear whether sub-micron MOS devices will be used for producing high-frequency analogue functions or whether an alternative technology will arise.

Finally, silicon designers should be aware of the many new devices that are rapidly being developed using III–V compound semiconductors based on GaAs. These will undoubtedly be used in circuits operating well beyond the limits of silicon, often in conjunction with optical signal processing. However, it remains to be seen whether they ever compete economically with silicon for general-purpose applications in digital and/or analogue electronics.

Appendix 1 Junction Capacitance

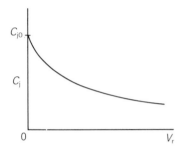

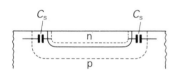

The reverse-biased p–n junction behaves electrically like a capacitor with a very small leakage current. The capacitance is due to the charge in the depletion layer that forms at the boundary between the n- and p-type regions. Since the width of the depletion layer increases with reverse voltage V_r the junction capacitance per unit area, defined as $C_j = dQ/dV_r$, falls as shown.

The value of C_j and the way it changes with reverse voltage depend on the doping concentration and profile of the junction. It is often fitted to a power law of the form

$$C_j = C_{j0}(1 + V_r/\psi)^{-n},$$

where C_{j0} is the zero-bias capacitance, ψ is the equilibrium potential difference across the junction, $0.6–0.7\,\text{V}$, and n is a parameter with a value between about 0.3 and 0.5. Typical values of C_{j0} for the junctions of MOS ICs are $0.1–0.5\,\text{fF}\,\mu\text{m}^{-2}$.

The junction capacitance is proportional to the area of the p–n junction. For a junction formed by doping a small area, the **sidewall capacitance** C_s should be included. This follows the same expression but with different values of C_{j0} and n because the doping profile is different in the sideways direction.

Appendix 2 d.c. Analysis of the CMOS Inverter

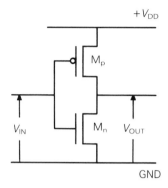

The CMOS inverter can be analysed graphically using the normal load-line method. The p-MOST, M_p, is regarded as the load for the n-MOST, M_n. However, the load line of M_p is not fixed as in a normal inverting amplifier because its characteristic changes with V_{IN}.

Figure A2.1 shows the characteristic curves of M_n and M_p for various values of V_{IN}. In each case the curve for M_p is drawn from right to left starting at V_{DD} on the horizontal axis. The intersection of the two curves at P gives the output voltage and steady-state current flowing through both transistors. We assume for simplicity that the numerical value of the threshold voltage is the same for both transistors.

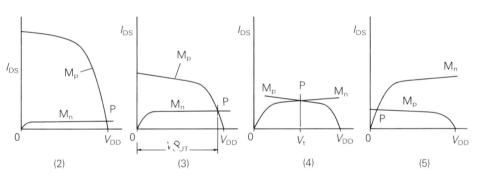

Fig. A2.1

As V_{IN} increases the circuit operates as follows:

1. V_{IN} less than V_T with M_n OFF, M_p ON and $V_{OUT} = V_{DD}$ (not shown in Fig. A2.1).
2. V_{IN} greater than V_T with M_n just conducting in its saturated mode while M_p is in the linear region.
3. V_{IN} approaching the transition voltage V_t. The current is greater than in 2 and V_{OUT} is falling more rapidly with V_{IN}.
4. $V_{IN} = V_t$. Both transistors saturated, current is a maximum, and V_{OUT} falls rapidly.
5. V_{IN} approaching $(V_{DD} - V_T)$. M_n conducts in its linear mode and M_p is saturated.
6. V_{IN} between $(V_{DD} - V_T)$ and V_{DD}. M_p is OFF, M_n is ON, and $V_{OUT} = 0$ (Not shown in Fig. A2.1).

The maximum slope of the transfer characteristic, at around V_t, depends on the saturation characteristics of the transistors. In the first-order model of the MOST, the current is constant in saturation and the slope is infinite.

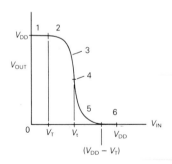

An infinite slope implies an infinite small-signal voltage gain at V_t if the circuit was used as an a.c. amplifier. With real transistors, particularly short-channel ones, the gain is reduced because the 'saturation' current rises with voltage.

The transition voltage can be found using the first-order model. The currents in the two transistors are

$$M_n, \qquad V_{GS} = V_t \qquad\qquad I_{DS} = \frac{\beta_n}{2}(V_t - V_T)^2 \qquad\qquad (A2.1)$$

$$M_p, \qquad V_{GS} = (V_{DD} - V_t) \qquad I_{DS} = \frac{\beta_p}{2}(V_{DD} - V_t - V_T)^2. \qquad (A2.2)$$

Equating (A2.1) and (A2.2), we find V_t to be

$$V_t = \frac{V_{DD} + V_T\left[\sqrt{\left(\frac{\beta_n}{\beta_p}\right)} - 1\right]}{\sqrt{\left(\frac{\beta_n}{\beta_p}\right)} + 1}.$$

In the electrically symmetrical case,

$$\frac{\beta_n}{\beta_p} = 1,$$

$$V_t = V_{DD}/2.$$

The maximum current taken by a CMOS inverter is found from either (A2.1) or (A2.2) by substituting back for V_t. In the electrically symmetrical case it is

$$I_{DD\,max} = \frac{\beta}{2}\left(\frac{V_{DD}}{2} - V_T\right)^2.$$

Appendix 3 Transient Analysis of the CMOS Inverter

Step Response

We consider the charging of the load capacitance C_L of an inverter when the input voltage is suddenly changed from V_{DD} to zero at $t = 0$ (Fig. A3.1).

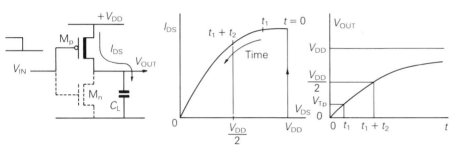

Fig. A3.1

The n-channel MOST, M_n, is turned off and C_L is charged by the drain current, I_{DS}, of M_p. The rising output voltage is determined by

$$I_{DS} = C_L \frac{dV_{OUT}}{dt}. \tag{A3.1}$$

To find the time for V_{OUT} to reach a certain value we integrate (A3.1)

$$t = C_L \int \frac{1}{I_{DS}} dV_{OUT}. \tag{A3.2}$$

The drain–source voltage of M_p is V_{DD} at $t = 0$ because C_L is initially uncharged, and the transistor is saturated, so that

$$I_{DS} = \frac{\beta_p}{2}(|V_{GS}| - |V_{Tp}|)^2, \tag{A3.3}$$

where $V_{GS} = V_{DD}$. After a time t_1, V_{OUT} will have risen to V_{Tp} after which M_p is no longer saturated and

$$I_{DS} = \beta_p V_{DS}(|V_{GS}| - |V_{Tp}| - |V_{DS}|/2), \tag{A3.4}$$

where $V_{DS} = V_{DD} - V_{OUT}$ and $V_{GS} = V_{DD}$ (still).

We want to calculate the delay time $(t_1 + t_2)$, where t_2 is the time for V_{OUT} to rise from V_{Tp} to $V_{DD}/2$. Time t_1 is found by substituting (A3.3) into (A3.2) and evaluating the integral between the limits $V_{OUT} = 0$ and $V_{OUT} = V_{Tp}$. This is a simple integral and the result is

$$t_1 = \frac{2C_L|V_{Tp}|}{\beta_p(V_{DD} - |V_{Tp}|)^2}. \tag{A3.5}$$

Time t_2 is found by substituting (A3.4) into (A3.2) and evaluating the integral between the limits V_{Tp} and $V_{DD}/2$. This is not simple but the result can be found by consulting a table of integrals and it is

$$t_2 = \frac{C_L}{\beta_p(V_{DD} - |V_{Tp}|)} \log_e \left(\frac{3V_{DD} - 4|V_{Tp}|}{V_{DD}}\right). \tag{A3.6}$$

In many CMOS processes $V_T = 0.2V_{DD}$ (e.g. $V_{DD} = 5\,V$, $V_T = 1\,V$). Substituting this into (A3.5) and (A3.6) gives

$$t_{pLH} = t_1 + t_2 = \frac{1.25C_L}{\beta_p V_{DD}} \left(\frac{1}{2} + \log_2 2.2\right)$$

or

$$t_{pLH} = \frac{1.62C_L}{\beta_p V_{DD}}. \tag{A3.7}$$

By changing the limits of the integrals the 10–90% risetime of V_{OUT} can similarly be shown to be

$$t_{LH} = \frac{3.7C_L}{\beta_p V_{DD}}. \tag{A3.8}$$

Ramp Response

In practice the input voltage always has a finite falltime t_{HL} which increases the delay from the value of (A3.7). This cannot be evaluated analytically, but computer simulation shows that the propagation delay increases to

$$t_{pLH} = \frac{1.62C_L}{\beta_p V_{DD}} + 0.257t_{HL} \tag{A3.9}$$

and, interestingly, that the output risetime t_{LH} (A3.8) is independent of the input risetime.

The transients in a chain of CMOS circuits of uniform size have the same risetimes and falltimes. Substituting $t_{LH} = t_{HL}$ from (A3.8) into (A3.9) therefore gives

$$t_{pd} = \frac{2.57C_L}{\beta_p V_{DD}}. \tag{A3.10}$$

A similar expression, but with β_n replacing β_p applies for the delay in discharging C_L.

Appendix 4 Bistables in CMOS Integrated Circuits

The transparent D-type (Fig. 4.21) is the basic element in most CMOS bistable circuits. Edge-triggered D-types are usually made with two transparent D-types in a master–slave arrangement, although a genuine edge-triggered D-type may have smaller set-up and hold times. The SR bistable is not recommended because the enabled D-type is a better alternative.

The J–K is a general-purpose bistable that is made in very large numbers as a standard-product IC. However, in an ASIC it is better to use whatever single-purpose bistable is required by the logic design. The J–K is a large circuit and its versatility is not normally needed in an IC.

J–K bistables are included in semi-custom cell libraries because some ASIC designs are direct copies of sub-systems built previously with standard-product ICs.

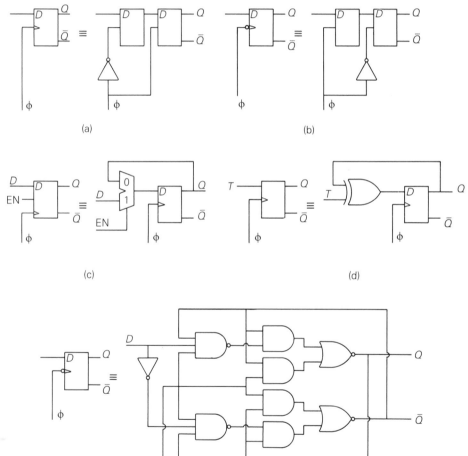

Fig. A4.1   (a) Master–slave positive-edge triggered D-type. (b) Master–slave negative-edge triggered D-type. (c) Enabled D-type. (d) T-type. (e) A genuine edge-triggered D-type.

Appendix 5 Digital and Analogue Hardware Functions

This appendix lists some of the electronic functions that are used at the block-diagram level of the design of electronic systems.

Digital Functions

Buffers, inverters.
Gates, AND, NAND, OR, NOR, XOR, XNOR. Tristate gates.
Bistables, D, T, JK, SR.
Switches, open-closed, multiplexers, demultiplexers.
Programmable logic arrays.
Finite-state machines, Moore, Mealey.
Memory, ROM, static and dynamic RAM, EPROM, EEPROM.
Registers, serial/parallel input and output.
Shift registers, barrel shifters.
Counters, any number base.
Decoders, coders, data selectors.
Adders, subtracters, comparators, arithmetic logic units.
Multipliers, dividers.
Timers.
A/D and D/A converters.

Analogue Functions

Amplifiers, single and differential input and output, op amps, wideband, tuned, d.c. or a.c. coupled.
Logarithmic and exponential amps.
Oscillators, sinusoidal, relaxation, voltage controlled.
Pulse generators, monostables, waveform generators.
Filters, low-, high- and band-pass, band-stop.
Analogue switches, multiplexers.
Voltage reference sources, current sources, voltage regulators.
Comparators, Schmitt triggers.
Analogue adders, subtracters, multipliers, dividers.
Timing circuits.
Zero-crossing detectors.
Rectifiers, detectors.
Modulators, demodulators.

Appendix 6 SPICE Level 1 MOST Parameters

Conduction Parameters

VTO   Threshold voltage for a long-channel transistor.

KP   Gain factor equal to β if $W/L = 1$ (page 54).

GAMMA   Coefficient giving the variation of threshold voltage with source–substrate voltage.

PHI   Surface potential of silicon when channel is formed.

LAMBDA   Parameter giving the slope of the saturated part of the MOST output characteristic.

As an alternative to entering the values of KP, GAMMA and PHI they will be calculated within SPICE if the following parameters are given:

TOX   Gate oxide thickness in metres.

UO   Low-field mobility of the carriers in the channel in square centimetres per volt per second.

NSUB   Substrate or well doping in centimetre^{-3}.

Several other parameters can be specified but they only have a small effect that is within the accuracy of Level 1 circuit modelling and they can be omitted.

Capacitive Parameters

The capacitances from the source and drain to the substrate will be evaluated in SPICE only if their areas, AS and AD (square metres), are added to the line describing each transistor in the circuit file. The capacitances of the sides of these junctions to the substrate will also be evaluated if the junction peripheries, PS and PD (in m), are given in the same way. The parameters used in calculating these capacitances as in Appendix 1 are:

CJ   Zero-bias capacitance C_{j0} of the junction in farads per square metre.

MJ   Parameter for evaluating the change of capacitance with voltage.

PB   Equilibrium voltage ψ across the junction.

CJSW   As CJ, but for the sides of the junction and in farads per metre.

MJSW   As MJ, but for the sides of the junction.

It is scarcely worth including these capacitances in Level 1 simulations as they are nearly always less than 15% of the total circuit capacitance.

References

IC Fabrication

Jaeger, R.G., *Introduction to Microelectronic Fabrication*, Addison-Wesley (1988).

Sangwine, S.J., *Electronic Components and Technology*, Van Nostrand Reinhold (1987).

Sze, S.M., *VLSI Technology* (2nd edn), McGraw-Hill (1988).

Devices

Navon, D.H., *Semiconductor Microdevices and Materials*, Holt, Reinhart and Winston (1986).

Sparkes, J.J., *Semiconductor Devices*, Van Nostrand Reinhold, UK (1987).

Digital Design

Barna, A. and Porat, D.I., *Integrated Circuits in Digital Electronics* (2nd edn), Wiley (1987).

Downton, A.C., *Computers and Microprocessors*, Van Nostrand Reinhold, UK (1984).

Gibson, J.R., *Electronic Logic Circuits* (2nd edn), Arnold (1986).

Green, D., *Modern Logic Design*, Addison-Wesley (1985).

Stonham, T.J., *Digital Logic Techniques* (2nd edn), Van Nostrand Reinhold, UK (1987).

Wilkins, B.R., *Testing Digital Circuits*, Van Nostrand Reinhold, UK (1986).

Circuits and ICs

Elmasry, M.I., *Digital Bipolar Integrated Circuits*, Wiley (1983).

Hodges, D.A. and Jackson, H.G., *Analysis and Design of Digital Integrated Circuits*, McGraw-Hill (1983).

Ritchie, G.J., *Transistor Circuit Techniques* (2nd edn), Van Nostrand Reinhold, UK (1987).

CMOS ICs

Allen, P.E. and Holberg, D.R., *CMOS Analogue Circuit Design*, Holt, Reinhart and Winston (1987).

Weste, N. and Eshraghian, K., *Principles of CMOS VLSI Design*, Addison-Wesley (1985).

IC Design

Jones, P.L. and Buckley, A. (eds), *Electronics Computer Aided Design*, Manchester University Press (1989).

Mead, C. and Conway, C.A., *Introduction to VLSI Systems*, Addison-Wesley (1980).

Naish, P. and Bishop, P., *Designing ASICs*, Ellis Horwood (1988).

Answers to Problems

Chapter 1

(a) 15 600 for L = 2 μm   127 500 for L = 0.7 μm
(b) 140 600 for L = 2 μm   1 148 000 for L = 0.7 μm

Chapter 2

(2.2) 1×10^{13}; (2.3) 1×10^{6}; (2.6) 53.5 fF

Chapter 3

Yes. With smaller chips the yield will increase to 30.2% and the output will rise by 405 000 chips per year. This will give an additional profit of £1 113 750 which is well above the redesign cost.

Chapter 4

(4.1) 4.9 fF, 71.5 μA V^{-2}; (4.2) 7.5 μm; (4.3) 1.4, 51, 173, 366, 631 μA; (4.4) 2.06 V; (4.5) 0.14 ns; (4.6) 0.84 ns; (4.7) 1.27 ns; (4.8) 89 μW; (4.9) 3.96 kΩ, 0.49 ns.

Index